AF564904

MICROBIAL PHYSIOLOGY

MICROBIAL PHYSIOLOGY

By

DR. S. SUNDARA RAJAN

Principal

Reva Institute for Science and Technology Studies

R.T. Nagar

Bangalore – 560 024

ANMOL PUBLICATIONS PVT. LTD.

NEW DELHI - 110 002 (INDIA)

ANMOL PUBLICATIONS PVT. LTD.
4374/4B, Ansari Road, Daryaganj
New Delhi - 110 002
Ph.: 3261597, 3278000
Visit us at: www.anmolbooks.com

Microbial Physiology

First Edition, 2003

ISBN 81-261-1333-2

PRINTED IN INDIA

Published by J.L. Kumar for Anmol Publications Pvt. Ltd., New Delhi - 110 002 and Printed at Mehra Offset Press, Delhi.

CONTENTS

PREFACE

Metabolic processes in living organisms in both microorganisms and higher organisms have been built upon a common foundation from where they (processes) have diversified depending on the evolutionary heirarchy.

Microbial Physiology deals with all the metabolic processes in microorganisms such as growth, culture, bioenergetics, bacterial photosynthesis, respiration etc. The book consists of six chapters and surveys all the parameters of microbial metabolism.

The book is meant to be a text book for students of life science, pharmacy, agriculture etc. at the graduate and post graduate level.

I am thankful to Mr. J.L. Kumar, M.D. of Anmol Publications Pvt. Ltd., New Delhi for publishing this book.

—**Dr. S. Sundara Rajan**
Principal
Reva Institute for Science and
Technology Studies, R.T. Nagar
Bangalore – 560 024

1

CULTURE OF MICROBES

Microbial sutdy involves various methodologies. Microbes present in natural conditions grow, mutliply and die. In the process, they affect our life in various ways both for the good as well as bad. Naturally, then we have to understand their life processes (microbes), as to how they are affected if certain requirements are not available. In natural conditions, microbes are rarely present in pure condition; there will always be mixed populations. Under such circumstances, it will be difficult to study them. In order to study the various life processes of microbes, they must be cultured on artificial media under laboratory conditions; isolated in pure form, and only then a meaningful study is possible.

A culture may be defined as growth of microbes under contamination free condition in the laboratory on a suitable medium.

The term culture is usually employed for an induced growth of microbes under *in vitro* (Latin; *vitro* - glass) conditions. Microorganisms as the very name indicates are extremely minute in size. It is difficult to obtain any information by the study of a single individual. Hence, microbes are always cultured to obtain populations of cells. Microbial populations are usually referred to as a **colony** (pl. Colonies). A colony develops from the multiplication of a single cell, and hence the population of cells are always homogenous. When a colony consists of only one type of cell it is called pure culture. In a mixed culture there will be cells of different types.

CULTURE MEDIUM

Microrganisms can be cultured in the laboratory under aseptic conditions using sterilized glass ware and other appliances. In order to cultivate the microbes, they should be provided nutrition on a medium in nearly as natural a way as possible. Preparation of suitable nutrient substances is the single most important factor in the cultivation of microbes in the laboratory. Nutrient substance prepared in a suitable manner (so that the microbes can grow on them), is referred to as the *culture medium* (Pl. media). Different microorganisms require different nutrient materials Thus there will be different types of media. Basically media may be classified into the following four types –

1. Natural (empirical) media
2. Semisynthetic media
3. Synthetic media and
4. Living media.

1. Natural (empirical) media

Natural media do not have any addition of specific nutrients. Natural food substances such as milk, vegetable juices, diluted blood, meat extracts etc. have been used since a long time by microbiologists for culturing microbes. This is also called an empirical medium because the exact chemical composition and the specific percentage of the chemical components is not known. Now a days peptone (digested protein rich in amino acids that provides nitrogen and carbon) is a major ingredient in natural media. The sources of peptone may be animals or plants, and in composition they (peptones) are highly variable. The chemical composition of peptones are partially known (both in concentration and identity). They are known to possess peptides, polypeptides, aminoacids, carbohydrates etc. in addition to many organic and inorganic nutrients. Peptones seem to provide the microbes in any easily digestible form P, S, N and C.

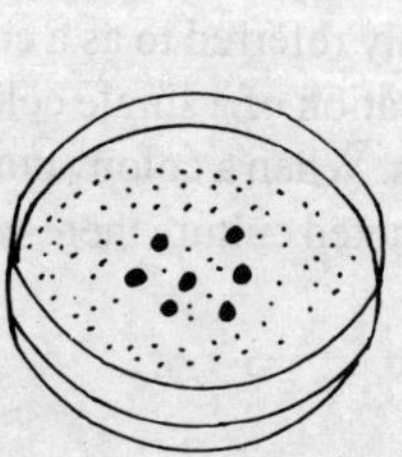
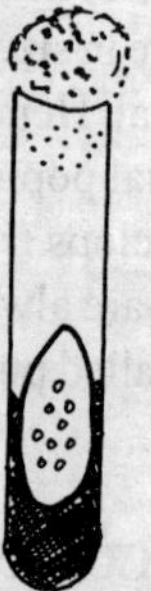

Fig. 1.1 Culture of Microbes

Method of culture, Petridish (left); Agar slant (centre); Broth (right)

Another important ingredient that is frequently used in natural media are the meat extracts and infusions. Beef extract and meat infusion (prepared by soaking fresh ground meat in water), are very commonly used in natural media. Beef extract is a rich source of minerals, organic micronutrients, proteins and carbohydrates. Yeast extract is often added to the beef extract to supplement vitamins. Beef extract may be used in liquid form when it is called *infusion broth* or *nutrient* or *nutrient solution.*

Many vegetable and fruit extracts are also used in natural media. These include extract of tomato, orange, apple etc. The choice of the extract depends on the type of microbe to be cultured. The natural medium with the addition of these extracts, then becomes a *special medium.*

Semisynthetic media

A semisynthetic medium is one in which the chemical composition (of the medium) is partially known. In other words, a semisynthetic medium has a natural medium (unknown chemical composition) to which certain specific nutrients (known chemical composition) are added. A medium which has agar becomes a semisynthetic medium. Some of the semisynthetic media are–porato dextrose agar (PDA), czapek dox agar, oat meal agar, beef peptone and nutrient agar etc.

Synthetic media

A synthetic medium is one in which all the chemical ingredients are mixed in definite proportions. They are usually prepared using chemically pure known inorganic and organic compounds. These media are also called **chemically defined media** as they have properly identified (defined) ingredients in specific proportions. Formation of a synthetic medium requires a prior understanding of proportions. Formation of a synthetic medium also requires a prior understanding of the nutritional requirement of the microbes to be cultured. Some of the synthetic media may have only inorganic components. For instance, *Thiobacillus Thiooxidans,* a soil inhabiting chemilithotroph can be cultured on a medium containing – (NH_4) $2SO_4$ 0.2G; Mg SO_4 $7H_2O$-0.5G; KH_2PO_4-3.0g; $CaCl_2$-0.25g; S (powdered) - 10.00g; distilled water in required quantity to make the volume to one litre.

In modern microbiological laboratories, dehydrated media (readily available) are used. These are available in ready to use mixtures which need to be weighed and added to a specific quantity of water.

A synthetic medium may be classified into two categories – General purpose medium and selective medium. A general purpose medium also known as enriched medium has standard ingredients known to help microbial growth, and can be used to culture a variety of microorganisms. A selective medium which has particularly selected ingredients is used for the culture of particular microbes only. Another kind of medium is differential medium used for biological assay of vitamins, antibiotics etc. This is called the assay medium. Some of the commonly used synthetic, media, are – Czapeck solution, Richards solutions, Ranlin's solution, Martins rose Bengal medium etc.

Living media

A living medium consists of living cells or tissues (even callus or organs), which are used for the culture of strictly parasitic organisms like viruses or rickettsiae which can not be cultured on a non living medium. For the cultivation of viruses normally chicks are used.

Other classifications of culture media

Media may also be divided into four categories on the basis of consistency ie., their physical state. These are–liquid media, semisolid media, liquefiable solid media and solid media.

Another classification is based on the application or function of the medium. Based on this, media are classified as follows -

1. Cultivation media
2. Storage media
3. Enrichment media
4. Differential media
5. Assay media and
6. Maintenance media

Basic ingredients of culture media

Media are of various categories based on the purpose for which they are used, and as such they contain a wide array of organic and inorganic ingredients in varying proportions. Often a microbiologist devises his own medium to suit a specific culture of microbes. There are some basic ingredients which, however, find a place in most of the media. Some of these are described below.

1. Protein hydrolysate

Protein extractions may be obtained from a wide variety of sources including both plants and animals. The following substances are used as sources of protein hydrolysates – meat, fish (fresh), caesin (milk protein), gelatin, keratin, yeast extract, soya meal, cotton seed, sunflower seed, groundnut meal etc.

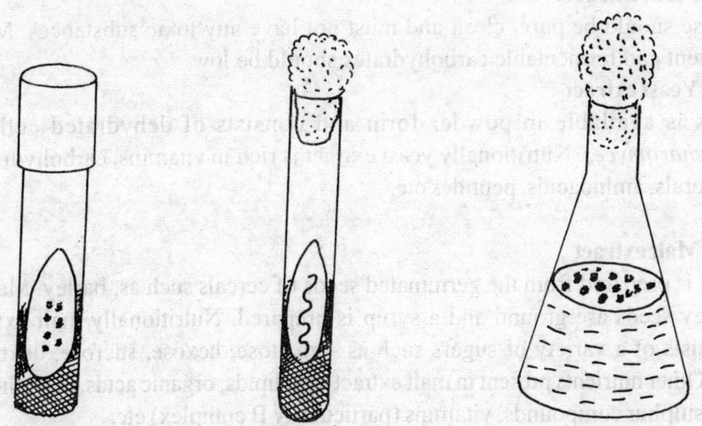

Fig. 1.2 Culture of Microbes
Methods of culture (Contd), Macartys slant (left); Streaking (centre); Flask (right)

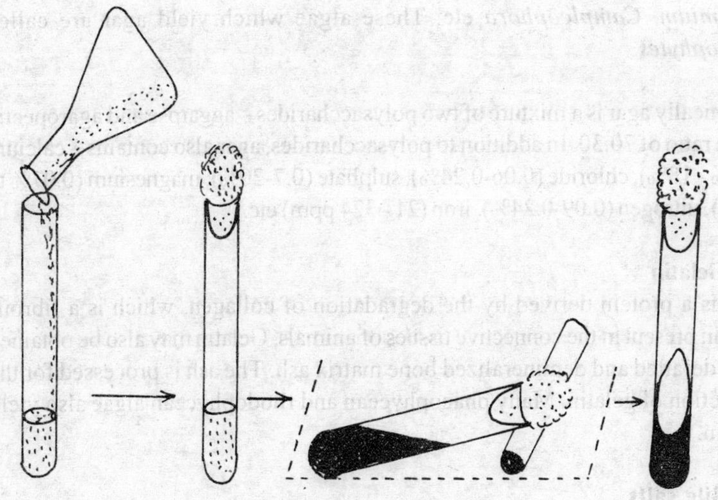

Fig. 1.3 Culture of Microbes
Preparation of Agar slant

2. Meat extracts

These should be pure, clean and must not have any toxic substances. Metal content and fermentable carbohydrates should be low.

3. Yeast extract

This is available in powder form and consists of dehydrated cells of *Saccharomyces.* Nutritionally yeast extract is rich in vitamins, carbohydrates, minerals, aminoacids, peptides etc.

4. Malt extract

This is prepared from the germinated seeds of cereals such as, barley. Malted barley seeds are ground and a syrup is prepared. Nutritionally malt extract consists of a variety of sugars such as – maltose, hexose, sucrose, dextrose etc. Other nutrients present in malt extract are– lipids, organic acids, phosphorus and sulphur compounds, vitamins (particularly B complex) etc.

5. Agar

Agar is a polysaccharide which can change from solid to liquid depending on the temperature. It has no nutrients but it is used for gelling the medium. It gets solidified below 40^{0}C. Agar (also called china grass) is obtained from a number of sea weeds belonging to the class *Rhodophyta*. Agar yielding red algae are – *Gelidium, Gracilaria, Hypnea, Chondrus, Pterocladia, Gelidiella, Gigartina, Ceramium, Campleophora* etc. These algae which yield agar are called *Agarophytes.*

Chemically agar is a mixture of two polysaccharides – aggarose and agaropectin in the ratio of 70:30. In addition to polysaccharides, agar also contains – calcium (o.8% - 18%), chloride (0.06-0.24%), sulphate (0.7-20%), magnesium (0.07% to 0.2%), nitrogen (0.09-0.24%), iron (21 - 324 ppm) etc.

6. Gelatin

This is a protein derived by the degradation of collagen, which is a fibrous protein present in the connective tissues of animals. Gelatin may also be obtained from defatted and demineralized bone matrix ash. The ash is processed for the extraction of gelatin. Many phaeophycean and rhodophycean algae also yield gelatin.

7. Bile salts

These are obtained from freshly collected bile of oxen. The components of bile liquid are – bile pigments, bile salts, mucin, proteins, lipids, enzymes and inorganic ions. Bile containing media are used in the culture of bile tolerant microorganisms.

8. Carbohydrates

A variety of simple sugars, disaccharides and complex carbohydrates are used in the culture media. The following are used in the media (Bhattacharya - 1986).

1. Pentose – arabinose, xylose, rhamnose etc.
2. Hexose - glucose, fructose, galactose etc.
3. Disaccharides - sucrose, maltose, lactose, trehalose etc.
4. Trisaccharides - raffianose etc.
5. Polysacharides - starch, insulin, dextrin, glycogen etc.
6. Alcohol - glycerol, crythritol, adonitol, mannitol, dulcitol, sorbitol etc.
7. Glucosides – salicin, aesculin.

Carbohydrates are prone to hydrolysis on heating and may also interact with salts which may retard the growth of microbes.

The following precautions will prevent the undesirable effects of carbohydrate decomposing (Bhattacharya, 1986).

1. Before autoclaving ,sugars should be completely dissolved.
2. Dextrose – Peptone media of high concentration should not be autoclaved.
3. The prescribed duration, pressure and temperature of autoclaving should be carefully adhered to.
4. pH should be maintained below 8.0.

9. Minerals

Some mineral elements have to be added for the culture of some microbes. These are K, Mg., Mn and SO_4. Some trace elements such as Cu, Co, Zn, Fe, Mg, molecules and Ca are also necessary.

10. Chelating agents

These are organic compounds which combine with heavy metals and keep them in solution in the media. Many substances such as – citrate, succinate, acelate etc. can complex with capsid, mg, fe etc. to prevent them from forming insoluble compounds. Certain aminoacids such as cysteine, histidine and glycine are also used as chelating agents. Presently, however, the best chelating agent used is EDTA (ethylene diaminotetraccetic acid).

TYPES OF CULTURES

Microorganisms can be cultured in media of various types depending on the requirement. The media used for culturing may be categorized into two depending on the consistency ie., solid or liquid. Solid media are prepared by adding either agar or gelatin to the nutrients in liquid, while liquid media do not have agar. The choice of the medium depends on the type of microbe to the cultured. The two types of cultures – liquid and solid are discussed below in some detail.

Liquid culture (liquid medium)

This is also known as the broth culture. A broth culture may be defined as a sterile liquid nutrient solution which does not have any solidifying medium (agar).

Broths are of different types

The choice depends on the requirement of microbe to be cultured. In most of the cultures two types of broth are used – **nutrient broth** and **glucose broth.** Broth cultures are ideal where growth performance of microbes are to be studied by techniques such as, turbidementric method. Turbidity or cloudiness in the nutrient solution indicates concentrated growth of microbes which are evenly dispersed in the medium. Turbidity may be measured in a spectrophotometer or colourimeter, where optical density indicates the measure of growth rate.

Broth cultures are also suitable for studying pellicle formation wherein, fungi cultured in liquid medium form large mats floating on the surface. Fungal nutritional studies also can be carried out using a broth culture. Study of the rate of sedimentation of bacterial cells is also possible using a broth culture.

Nutrient broth and glucose broth are two of the general liquid media used for the cultivation of a variety of bacteria. The following is the method used for the preparation of broth culture.

Requirement

1. *Nutrient broth*
   a. Peptone 5.0g
   b. Beef extract 3.0g (pH 7.0)
   c. Distilled water 1000ml
2. *Glucose broth*
   a. Peptone 10.0g
   b. Glucose 5.0g
   c. Sodium chloride 5.0g (pH 7.3)
   d. Distilled water 1000ml

In addition to the above, the following are also required –
1NHC1, 1Naoh, pH meter, distilled water, hot plate, autoclave, culture tubes, flasks etc.

Procedure

Nutrient broth. Put the weighed amount of peptone (5g), and beef extract (3g), in 500ml of distilled water. Heat the mixture and agitate with glass rods to dissolve the constituents. Add distilled water to make the volume to one litre. Adjust the pH of the medium to 7.0 either by adding acid or alkali depending on the necessity. Take test tubes and pour 10ml each. Conical flasks also may be used instead of test tubes, in which case pour 25ml to each flask. Apply cotton plugs. Autoclave at 121^{0}C under 151b pressure for 15 minutes. Allow the autoclave to cool. Remove the broth tubes (flasks), and store at room temperature for use.

Glucose broth may be prepared in the same fashion as the nutrient broth.

Shake cultures

These are nothing but liquid cultures. The culture flasks contain a large quantity of medium. Many a time the nutrients settle down to the bottom of the flask. As a result nutrients are not uniformly distributed all over the medium. This will affect the growth of the microbes and they do not develop properly. In order to overcome this, the culture flasks have to be agitated or shaken frequently to allow uniform distribution of nutrients. Instead of shaking the flask manually, this is a platform on which are placed the flasks. The flask shaker has an electric motor which shakes the platform and the flasks kept on it.

Solid cultures

In liquid cultures, the microbes are not fixed to the place of their origin and colonies cannot be studied or separated as they (microbes) are floating. This can be overcome by the use of solid media. Solid media can be prepared by the addition of agar to the medium which solidifies at room temperature. The solid surface of the medium is ideal for the spores or cells of microbes as they get trapped and grow very well. A colony originates from a single cell and it cannot shift to any other place.

Solid medium can be used either in test tubes or in petriplates. Accordingly, the solid media may be classified into agar plates (in petri plates), and agar slants (in test tubes).

Preparation of agar plates

1. Clean petriplates and wipe them dry.
2. Prepare the culture medium (including agar which melts on heating to 96^{0}C and hardens into a jelly on cooling to 40-45^{0}C.)
3. Carefully open the petriplate and pour the medium into the petriplate and close the lid.
4. Conduct the operations near a flame.

5. After pouring the medium, tilt the plate back and forth to allow for even distribution.
6. Allow the medium to settle and solidify.
7. After the medium gets solidified, keep them in an inverted fashion in an incubator to prevent the condensation of vapour into water drops.

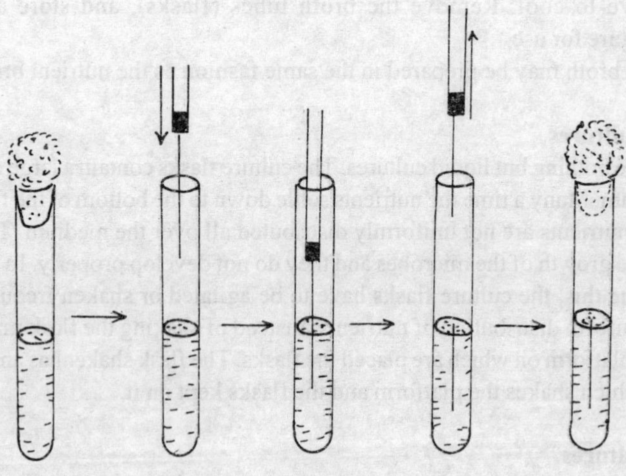

Fig. 1.4 Culture of Microbes
Agar stab culture

Preparation of agar slants

Agar slants are usually prepared in tubes. After the medium (about 10-20 ml) is poured into tubes, they are autoclaved. After removing from the autoclave the tubes are kept in a slanting position and the medium is allowed to solidify. After hardening, the surface of the medium will be slanting. The purpose of slanting is to increase the superficial area for the growth of the microbes, otherwise the surface would be as wide as the diameter of the tube.

Agar slants are ideally suited to study the cultural characters of microorganisms. Slants are also used for maintenance and transfer of cultures.

Agar stabs

Here also, the media is poured into the tubes, but instead of keeping the tubes in a slanting position, they are kept straight. As a result the surface of the solidified medium is even. Stab cultures are used to study the O_2 requirement of microbes, motility of microbes and to preserve the culture for a longer period.

Specific media for particular microbes
The description of the media given above is a general purpose one routinely used for the culture of microbes. However, not all microorganisms can be cultured on these media. Hence, special types of media are to be prepared, and particular nutrients needed have to be added to facilitate the growth of microbes. Some of these media are described below.

Nutrient agar and broth
This medium is used for the isolation of bacteria and actinomycetes. The components of the media are as follows.

a. Peptone 5.0g
b. Beef extract 3.0g
c. Nacl 5.0g
d. Distilled water 1000ml
e. Agar 15g

(Note : for nutrient broth, omit agar)
In case of nutrient broth the final volume should be made up to 1 litre. pH to be adjusted to 7.0.

Potato Dextrose Agar (PDA) medium
This medium is a typical solid medium used for the culture of a number of bacteria and fungi. The medium may be used in petriplates, test tube slants or test tube stabs. PDA is highly useful for the isolation of plant pathogenic fungi. Acidifying the medium with a few drops (for every 100ml melted agar) of 25% lactic acid is highly suitable for growth plant pathogens. The following is a detailed procedure for the prepartion of PDA.

Requirements
Peeled potato tubers (200g), dextrose (20g), agar (20g), corning beakers, potato peeler, knife etc.

Procedure
1. Take 500ml of water in a 5 litre beaker.
2. Peel the skin of potato; cut into small pieces and boil in water for about 30 minutes to make a thick paste.
3. Add some water to the paste and filter through a cheese cloth; squeeze out all the liquid.
4. Take 500ml of distilled water in a beaker and heat it.

5. Dissolve agar in water by gently stirring.
6. Mix the potato extract with agar solution.
7. Add water to make up the total volume to1 litre.
8. Distribute the nutrient liquid to either conical flasks or test tubes (quantity), nutrient liquid in each flask/tube depends on the requirement.
9. Cover the flask/tubes with cotton plugs.
10. Autoclave at 121°C or on 15 minutes under 151b pressure.
11. After removing from the autoclave, the tubes may be kept in slanting position or upright (for deep agar tubes).
12. After solidifying, the medium may be used for inoculation or may be stored under refrigeration.
13. If agar plates are required, the medium may be poured into petriplates under strict aseptic conditions and allowed to solidify before using for cultures.

Yeast extract mannitol agar

The composition of this medium is as follows

Mannital	10.0g
Dipotassium Hydrogen phosphate	0.5g
Magnesium sulphate	$7H_2O$ 0.2g
Sodium chloride	1.0g
Yeast extract	1.0g
Congo red (1% solution)	2.5ml
Agar	20.0g
Distilled water	100ml

Martin's medium

It is also called Martin's Rose bengal streptomycin agar. The composition of the medium is as follows.

Dextrose	10.0g
Peptone	5.0g
Potassium dihydrogen phosphatge	1.0g
Magnesium sulphate	1.0g
Rose bengal (one part in 30,000 parts of the medium)	
Agar	20.0g
Streptomycin	30.0g
Distilled water	100ml

Streptomycin vial has to be opened aseptically only at the time of preparation of medium. To one gram of streptomycin and 100 ml of double distilled water. Add 0.31 of this solution to each 100 ml of the basal rose bengal medium after it is cooled to 45°C.

Selective and enrichment media

These are the media that are specially meant to culture specific microbes, and are not generally used for routine cultures.

A selective medium is one which permits and promotes the growth of some specific group or type of organisms while preventing or retarding the growth of other microbes.

An enrich medium helps in the increased growth of only one kind of a microbe. In other words, an enrichment medium 'enriches' (increases the concentration) the microbial content in the medium.

Mannital salt agar medium

This is a typical example of a selective medium. As has been pointed out already a selective medium not only promotes the growth of specific microbes but also retards the growth of others and thus it helps in the isolation of bacteria. Selective action of the medium is due to the addition of certain chemicals to the medium. For instance, addition of crystal violet dye selectively inhibits the growth of gram positive bacteria and is added to a medium to culture enteric bacteria (gram negative bacteria). In the same way, addition of sodium azide at specific concentration into the medium will help in isolating lactic acid bacteria. There are many dyes used for selectively promoting the growth of *Pseudomonas* (cetrimide), halophytic bacteria (tetrazolium) etc.

Mannital salt agar is used for the culture of pathogenic stayphylococci, as most other bacteria are inhibited by the high concentration of sodium chloride in the medium. The fermenting ability of the staphylococci colonies induced by a yellow halo can be detected by mannitol present in the medium. Mannitol is metabolised and fermented by staphylococci and the acid produced can be detected by phenol red, a - pH indicator. The following is the composition and procedure for the preparation of the medium.

Mannital	10.0g
Beef extract	1.0g
Peptone	10.0g
Sodium chloride	75.0g
Phenol red	0.025g
Agar	15.0g
Distilled water	100ml

Procedure

Dissolve the components (beef extract, peptone, sodium chloride, mannitol, phenol red) in 500ml of distilled water and add agar solution (dissolve 15g, agar in 500ml of water). Make the total volume to one litre. Pour the medium into suitable culture tubes/ conical flasks and autoclave for 15 minutes under 151b pressure.

After the medium is sterilized and sufficiently cooled, inoculate the medium with *Staphylococcus aureus*. If necessary pour the medium into petridishes for the purpose of culturing. Incubate the inoculated plates/tubes at 37^{0}C for 48hrs and observe the growth of bacteria.

Enrichment medium

Enrichment or enriched medium is one in which the nutritional components are balanced in such a way as to selectively enhance the growth of a certain microbe in a mixed population of the inoculum. This method is suitable for those microbes which are in extremely small numbers in the sample, and whose growth is slower than the others. The basis behind an enrichment medium is perfect regulation of nutrients, and culture conditions such as temperature, oxygen, light, pH etc. in such a way that it provides optimal environment and promotes the growth of the required microbe. For example, if a medium containing salt solution with $NaNO_2$ (pH 8.5) is inoculated with a sample from the garden soil and is incubated in the dark (in the presence of air) at a temperature of 25-30^{0}C, pure colonies of *Nitrobacter* develop. Obviously even though garden soil sample consists of other microbes, given the culture conditions, they will not grow. Acterotrophic bacteria can be enriched by the addition of plant or animal extracts into nutrient broth or agar.

Enrichment medium is ideal for the isolation and purification of specific microbes.

Differential medium

A differential medium is one which will cause certain colonies to develop differentially from others by producing a characteristic change in the microbial growth or in the medium surrounding the colonies. Differential medium is employed to determine differential reactions which will permit presumptive medium. It can be used to differentiate hemolytic properties of streptococci by their reactions on the medium. Thus streptococci may be distinguished in a mixed culture. Red blood cells get hemolyzed around the colonies of streptococci and form a clear zone. Such a clear zone is not formed around non hemoloytic bacterial colonies. Hemolytic zones are classified into alpha, beta and gamma corresponding to a greenish halo, a clear zone and no reaction respectively. The constituents of blood agar medium are as follows.

Infusion from beef heart 500.0g

Tryptose	5.0g
Sodium chloride	5.0g
Agar	15.0g
Distilled water	100ml

Dissolve the ingredients given above except agar in 500ml of distilled water, and add it to agar solution (15 gms of agar dissolved in 500ml of distilled water). Autoclave the medium for about 15 minutes under 151b pressure. Cool the sterile defibrinated blood and mix gently but thoroughly. Pour the medium into suitable culture plates. Inoculate with *Streptococcus faecallis* and incubate at 37°C for 48hrs. Observe the hemolytic reaction of bacterial colonies.

The following zones are seen –

(a) Greenish discolouration of the medium surrounding the bacterial colonies (α hemolysis) due to the breakdown of RBCs with the reduction of hemoglobin to methemoglobin.

(b) Clearzone surrounding the bacterial colonies (ß hemolysi) with complete destruction of RBCs and use of hemoglobin by bacteria.

(c) No reaction zone on the medium (γ hemolysis neither discolouration nor a clear zone) indicating absence of hemolysis.

INOCULATION

Artificial induction of microbes into the culture medium is called inoculation. Inoculation is the most important technique for facilitating a proper growth of microorganisms under aseptic conditions.

Microorganisms from the original stock culture are to be introduced into either petriplates or agar slants prepared in culture tubes (for preparation of agar slants see earlier in the same chapter). The procedure of inoculation into an agar slant culture is as follows.

1. Hold the tube containing the stock culture and the tube containing the culture medium in the left hand. Hold the inoculating needle or plantium loop in the right hand.
2. Hold the needle/loop on to a flame of bunsen burner until it becomes red.

3. Remove the cotton plugs of both the tubes with the finger of the right hand, and simultaneously expose the mouth of the tubes to the flame device.

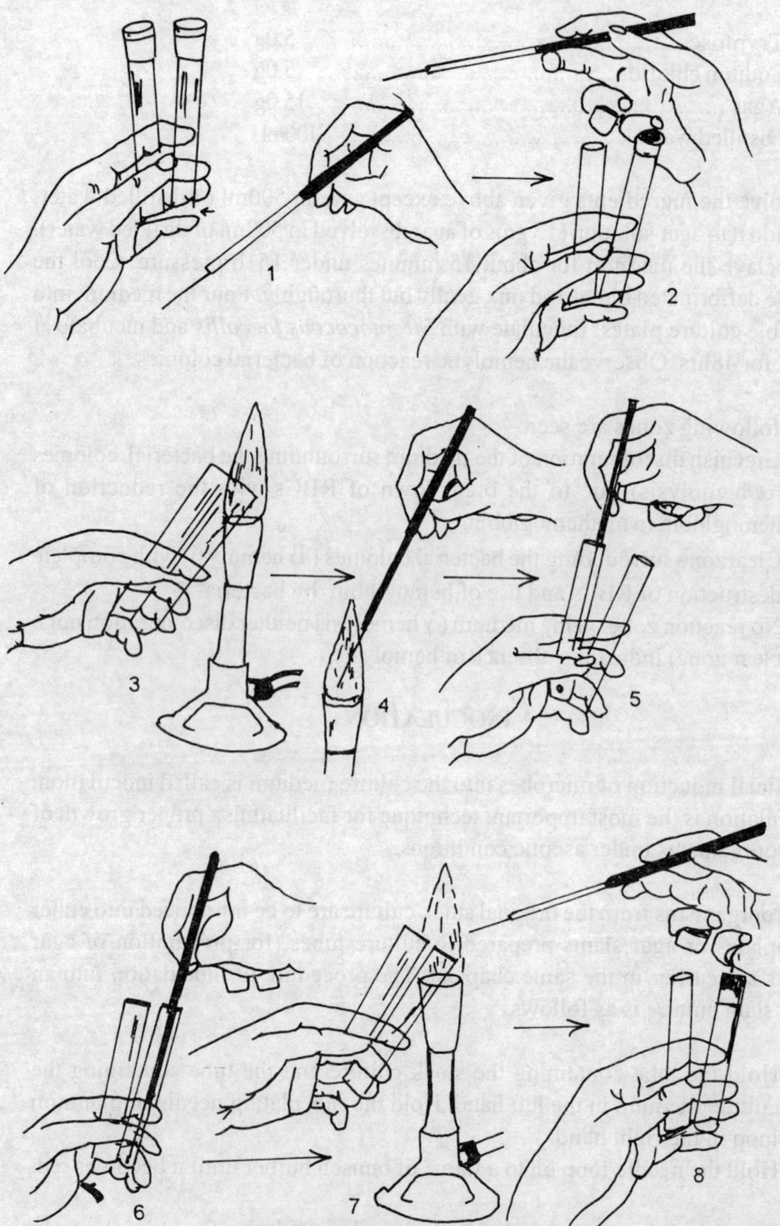

Fig. 1.5 Culture of Microbes
Procedure for inoculation

4. Immediately after the removal of the plug and flaming the mouth, flame the loop/needle again and dip it into the tube containing the stock culture.
5. Take out a little quantity of the inoculum (from the stock culture), and immediately introduce it into the other tube containing the culture medium. The inoculum should be rubbed on to the surface of the agar medium.
6. Flame the needle and the mouths of both the tubes again once or twice.
7. Plug the tubes immediately and transfer the tubes for incubation under suitable conditions.

Precautions

(a) All steps particularly after removing the cotton plugs are to be carried out as quickly as possible to prevent contamination.

(b) After removing from the tube, till replacement, cotton plugs should be held in the hand only and *must not be kept on a table or anywhere.*

PURE CULTURE TECHNIQUES

Microbes, in nature, are never found in pure populations. In the source – water, air or soil, microbes are always found as a mixed population. Even in the infected plants or animals, the affected parts harbour a variety of microorganisms.

In the laboratory study of microorganisms, a mixed population is of a disadvantage. For instance, a specific microbe on which some experiments have to be carried out may be in very small numbers. In such an instance if one has to chemically analyse a metabolic product produced by that specific microbe it will be practically impossible as the quantity is very less. Further, in a mixed population one group of microbes may adversely influence the function of another. Hence, any meaningful work on any microbe is well nigh impossible in a mixed population. The solution for this is to obtain pure population of required microbes. The advantages of pure populations (colonies) of microbesin, a laboratory study are –

1. Growth conditions/requirements may be clearly studied.
2. Metabolic products may be analysed.
3. Large populations of pure colonies are available for genetic studies.

Isolation of a single microbe for the purpose of pure culture is easier said than done. A great variety of laboratory techniques are to be involved to isolate required microbes from natural habitats. Culturing of microbes from natural populations with diverse microbes will often lead to overlapping of colonies on the medium making it difficult to separate them. Hence, it becomes necessary to follow methods which allow colonies to develop distinctly and separated from each other to permit easy manipulation. Colonies which are distinct tend to be pure and may be picked up for further culturing.

Several laboratory techniques have been devised to obtain a pure culture from mixed culture developed from natural populations. It must be pointed out however, that liquid media by their very nature are unsuitable for pure cultures. Colonies grown in liquid media can never be fixed and localized. Solid agar media are most suitable because, they can trap and fix a cell or a spore of microbe, and lead to the development of a colony in the place of original location.

The techniques to obtain pure cultures may be classified as follows.

1. **Common methods**
   a. Pure plate method
   b. Streak plate method
   c. Spread plate method
   d. Serial dilution method

2. **Special methods**
   a. Single cell isolation
      i. Capillary pippette method
      ii. Micromanipulator method
   b. Enrichment method
   c. Selective media method
   d. Differential media method

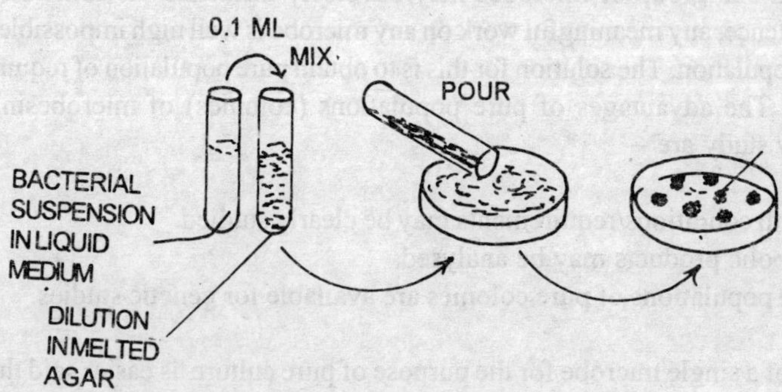

Fig. 1.6 Culture of Microbes
Pour plate technique

Pure plate method : The principle involved in this method is to dilute the sample inoculum mixed with the melted agar medium. This allows for a thorough and uniform distribution of cells or spores of microbes within the medium. The medium is then poured into peptridishes, allowed to solidify and incubated under suitable conditions. As a result of this dilution discreet and well separated colonies develop in the culture media. As the inoculum is distributed evenly in the medium both surface as well as subsurface (within the agar medium), colonies develop in the medium.

Procedure of pure plate method

(a) Take four petridishes; sterilize them properly in a hot air oven. Remove them and mark the numbers 1 to 4 with the help of a glaze marker.
(b) Prepare four tubes of nutrient agar with each tube having 12-15ml of medium. Cool the tubes to 45°C and number them (1 to 4) (see earlier in the same chapter for the details of medium).
(c) Prepare a suspension of the source material such as soil sample, sewage or bacterial inoculum etc. in 20ml of sterile distilled water.
(d) Transfer the microbial suspension into the four tubes (marked 1-4), containing nutrient agar medium as follows.

I. Tube

1. Transfer one loopful of the medium, stir the loop in the medium. Gently roll the tube between the palms to facilitate even dispersion of the inoculum in the medium. Keep the tube in water bath for incubation at 45°C.

II. Tube

2. Collect two loopful of inoculum (mixed in the medium) from tube number one and transfer it to tube no two. Follow the same procedure to evenly distribute the inoculum (as in tube one). Keep the tube for incubation in water bath.

III. Tube

3. Transfer two loopful of inoculum from tube no 2 to tube no 3. Follow the same procedure and proceed for incubation.

IV. Tube

4. This will be a control with no microbial inoculum.

Transfer of cultures to petriplates

1. Arrange the petriplates serially 1 to 4.
2. Take out the incubated tube no. 1 from water broth. Wipe the outersurface perfectly dry. Carefully remove the cotton plug, flame the mouth of the tube and carefully pour the contents into petriplate no. 1 by raising the lid

slightly. Immediately close the lid to prevent any contamination. Gently rotate the petriplate to allow even distribution of the medium.

3. Similarly pour the contents of tubes 2,3 and 4 to petriplated 2,3 and 4 respectively.
4. After the medium solidifes, place the petriplates in an incubator at 30°C in an inverted position for 48hrs.

Note : The petriplates are to be kept in inverted position (ie., the lid should be below), in order to prevent the condensed water drops from falling back into the medium.

Observation

Colonies appearing in plate no.1 are not distinct, small in extent and overlap with one another. In contrast to this, in plate 2 the colonies are large, discreet and generally do not mix with other coloneis. Plate number 3 will have much larger and sparsely distributed (no crowding) colonies with distinct shape, colour, texture, size and form. This is due to the fact that from plate 1 to 3, there is progressive dilution of the inoculum and only a less number of microbes are allowed to grow on plate 3. Plate no 4 will have no microbial growth as there was no inoculum in tube no 4. In plate 2, and more particularly in plate 3, colonies with diverse characters will develop both on the surface and subsurface of the medium.

Streak plate method

In this method, a small amount of mixed (concentrated) inoculum is placed on the solid agar medium (in petriplate) with the help of a needle/loop. From the point of origin, thin and gentle lines of the needle/loop are drawn over the surface of the medium.

The purpose of streaking is to thin out the inoculum successively so that microbes get separated. A second plate also may be streaked from the same loop/needle without reinoculation. In the streak, microbes are crowded and colonies develop closely, but as the streaking proceeds, cells get separated as the needle contains fewer and fewer cells. Hence, at the last stretch of the streak few, well developed and clearly separated colonies will develop. It will be thus possible to obtain pure cultures. Various designs of streaking can be made on the medium as shown in figures. The following is a generalized procedure used in streak plate method.

1. Melt the autoclaved nutrient agar medium in a boiling water bath and cool to 45°C.

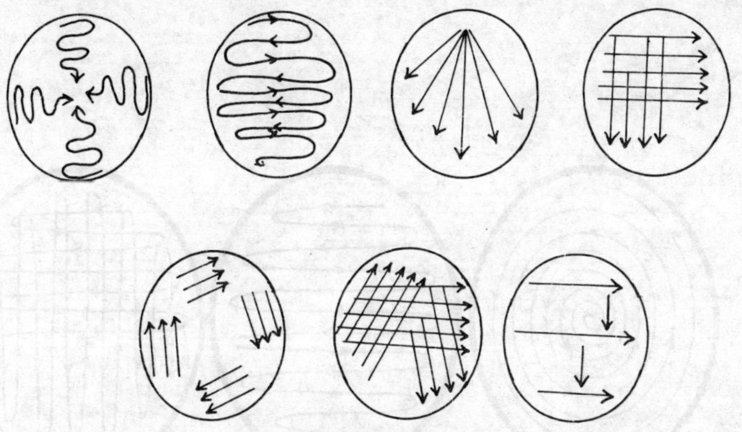

Fig. 1.7 Culture of Microbes
Methods of streaking

2. Pour the medium (about 15ml), into several petriplates carefully. Rotate the petriplates to allow for uniform distribution of the medium. Allow the medium to solidify. Strict aseptic conditions are to be followed to prevent contamination.
3. Take a small amount of bacterial suspension on to the needle or loop. Carefully lift slightly the lid and streak on the medium in any fashion as shown in the figure. Different designs of streaking mayh be done on different petriplates using the same sample source. Do not take a suspension a second time while streaking a plate.
4. Incubate the petriplates according the procedure at 30^{0}C for 48hrs.

Observation

Observe the different petriplates. It will be noticed that well developed and discrete colonies have developed at the end of streaks. The effect of different designs of streaking on the growth and separation of colonies may also be observed. Colonies that are clearly may further be transfered to agar slants/ petriplates to grow as pure cultures.

Sub culturing

Transfer of microbial colonies from one culture to another for the purpose of isolation and purification is called subculturing. Great care has to be taken during transfer to prevent contamination. Bacterial cultures kept in the same medium for a long time will not survive. Ideally, cultures are to be transfered to fresh media at an interval of 15 - 20 days.

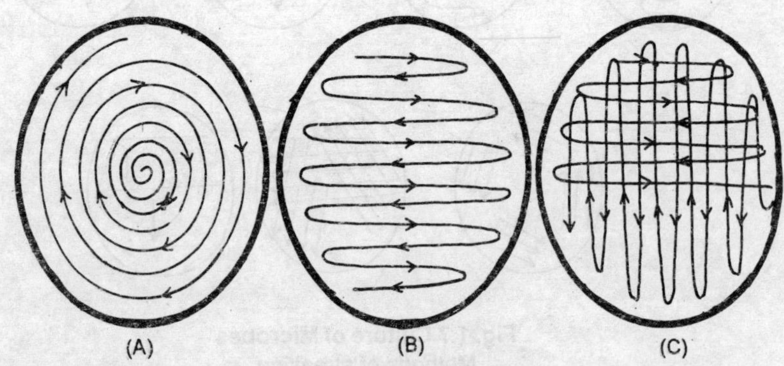

Fig. 1.8 Culture of Microbes
Methods of streaking (contd)

Spread plate method

In this method, an aliquot of the inoculum is placed on the medium surface and is spread uniformly with a bent sterile glass rod. This will help in diluting the cell concentration in the inoculum. By this method cell aggregates get separated. When colonies develop, they may be picked up and further spread on a fresh medium. Several such spreads may be needed before obtaining the pure cultures of diplococci, streptococci and staphylocci.

Serial dilution method

Among the methods used for purifying cultures, this is by far the simplest and most widely used method. In this method the original inoculum is subjected to serial successive dilutions, so that the concentration of the microbes (in a fixed quantity of the liquid), gradually becomes less and less. When these serial dilutions are plated, colonies will appear discrete and far removed from one another.

The principle of serial dilution may be explained as follows. If we prepare a 10ml sample of soil suspension and assume that it contains 1000 microbes, then each ml of this solution consists of 100 microbes. Take one ml of this and add 9ml of sterile distilled water to make the volume to 10ml. This dilution will have 1:10 or 100 microbes per 10ml, or 10 microbes per one ml. This may be called as 10^{-1} concentration of the original sample. From this dilution if we take one ml and add 9ml of sterile distilled water to make it to 10ml, then each ml will have one microbe. This concentration is 1:100 or 10^{-2} of the original concentration. This dilution of one microbe per ml is the starting point of pure culture. Further culturing will yield microbes of one type otherwise known as pure culture.

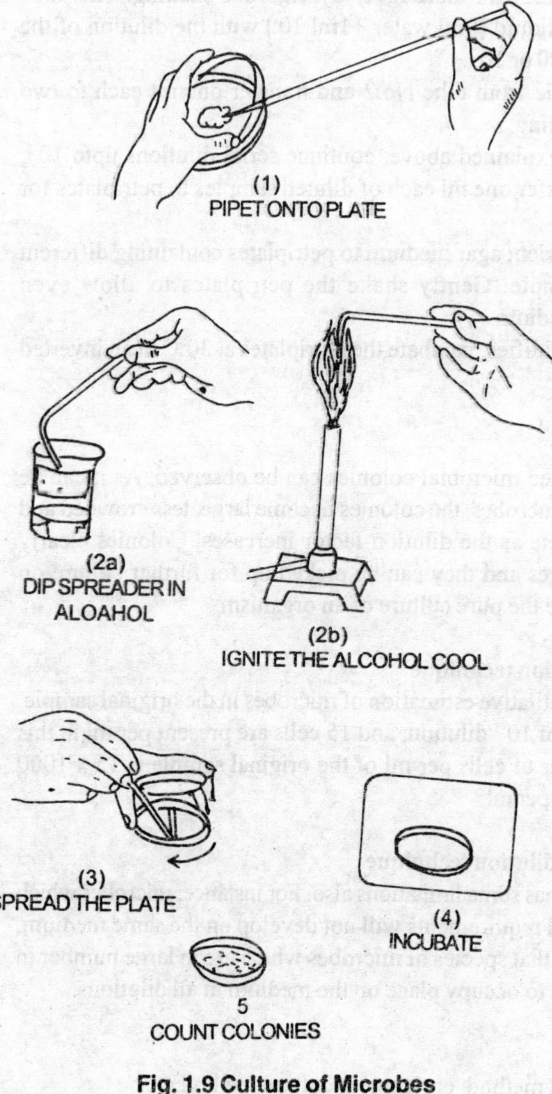

Fig. 1.9 Culture of Microbes
Spread plate technique

Procedure for serial dilution (A sample experiment)

1. Take 6 graduated sterile pipettes of 2ml each and also 12 sterilized petridishes.
2. Take 6 test tubes each having 9ml of sterile distilled water. Arrange them serially after marking the numbers 1 to 6.
3. Prepare nutrient agar medium, melt it; cool to 45°C and keep in water bath at 45°C.
4. With the help of a sterile pipetee take one ml or one gm of soil or sewage or any source inoculum and transfer to tube no 1. The tube now contains 10ml of liquid i.e., 9ml of water and one ml of inoculum. In otherwords, the inoculum is diluted 10times (10^{-1}). Mark the dilution factor on the tube.

5. Using another sterile pipettee take 2ml of the diluted inoculum (10^{-1}) from tube No.1 and trasfer one ml each to two petridishes. All transfers are to be done following strict aseptic conditions.
6. Using the same pipettee collect one ml of sample from tube no 1 and transfer it to tube no.2. Mix thoroughly by vigorous shaking. This tube now will have 10ml liquid (9ml water +1ml 10^{-}) with the dilution of the inoculum being 1/1000 or 10^{-2}.
7. Collect 2ml of sample from tube No.2 and transfer one ml each to two petridishes for culturing.
8. In the same way as explained above, continue serial dilutions upto 10^{-5}, and accordingly transfer one ml each of diluted samples to petriplates for further growth.
9. Add 7.5ml of the nutrient agar medium to petriplates containing different dilutions of the sample. Gently shake the petriplates to allow even distribution of the medium.
10. After the medium solidifies, incubate the petriplates at 30°C in an inverted position.

Observation

Afteer 48hrs the developed microbial colonies can be observed. As it can be seen from the growth of microbes, the colonies become large, less crowded and finally completely discrete as the dilution factor increases. Colonies clearly separated are pure cultures and they can be picked up for further culture on agar slants. This will give the pure culture of an organism.

Advangate of serial dilution technique

This can be used for qunatitative estimation of microbes in the original sample. Suppose we use one ml of 10^{-3} dilution, and 15 cells are present per ml in this dilution, the total number of cells per ml of the original sample is 15 x 1000 (dilution factor) or 15000 per ml.

Disadvantages of serial dilution technique

Serial dilution technique has some limitations also. For instance, microbes which have different nutritional requirements will not develop on the same medium. Another problem here is that species of microbes which are in large number in the original sample, tend to occupy place on the medium in all dilutions.

Special methods

These include single cell method, enrichment culture method etc.

Single cell method. In this, a single microbial cell is picked up from the culture medium and is allowed to grow. Two methods are employed to isolate single cells. These are:

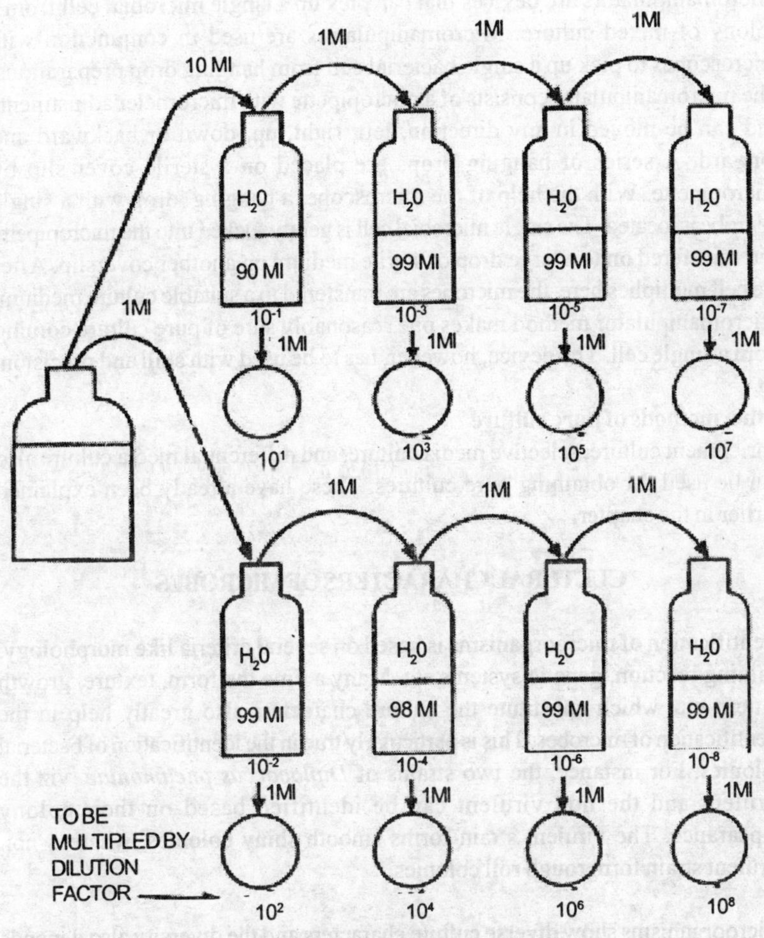

Fig. 1.10 Culture of Microbes
Serial dilution technique

1. Capillary pipette method

Several droplets of a suitably diluted culture medium are placed on a sterile glass cover slip using a sterile capillary pipette. Each drop of the medium is observed under microscope. A drop which contains only one microbe is removed through a capillary pipette and inoculated on to a fresh medium. The individual microbial cell multiples and produces a colony of pure culture.

2. Micromanipulator method

Micromanipulators are devices that can pick up a single microbial cell from a colony of mixed culture. Micromanipulators are used in conjunction with microscopes to pick up a single bacterial cell from hanging drop preparations. The micromanipulator consists of a micropipette with micrometer adjustments and can be moved in any direction, left, right, up, down or backward and forward. A series of hanging drops are placed on a sterile cover slip by micropipette. With the help of the microscope, a hanging drop with a single microbe is located. The single microbial cell is gently sucked into the micropipette and transfered on to a large drop of sterile medium on another coverslip. After the cell multiplies here, the microbes are transfered to a suitable culture medium. Micromanipulator method makes one reasonably sure of pure culture coming from a single cell. The device, however, has to be used with skill and precision.

Othermethods of pure culture

Enrichment culture, selective media culture, and differential media culture also can be used for obtaining pure cultures. These have already been explained earlier in the chapter.

CULTURAL CHARACTERS OF MICROBES

Identification of microorganisms is based on several criteria like morphology, staining reaction, genetic systems etc. Many a time the form, texture, growth pattern etc. which constitute the culture characters also greatly help in the identification of microbes. This is particularly true in the identification of bacterial colonies. For instance, the two strains of *Diplococcus pneumoniae,* viz the virulent and the non virulent can be identified based on their colony appearance. The virulent strain forms smooth shiny colonies, whereas non virulent strain form rough roll colonies.

Microorganisms show diverse culture characters and the diversity also depends on the type of medium used for culturing. Growth parameters such as size, colour, texture, margin, elevation, consistency etc., are of value in the identification of colonies. However, not all the above characters can be studied on all the media. While the above mentioned characters are studied in solid media, the amount of growth and distribution can be studied in liquid cultures such as nutrient broth. We will study below in detail the various cultural characters when grown in different types of media.

1. Colonies on agar plates

Following characters of colonies can be studied on agar plates.

A. Size

The size of the colonies varies from very small (pinpoint), measuring only a mm or less to large colonies ranging between 5–10mm in diameter. While using size as a criterion one should take into consideration only those colonies which are situtated far apart. Colonies situated close by tend to be small due to competition for nutrition. Further, the colonies which have grown for sufficient period only should be measured, as young colonies tend to be smaller in size. Unusually large colonies are encountered in bacteria like *Protes* and *Bacillus* where they spread on the entire medium on petriplates.

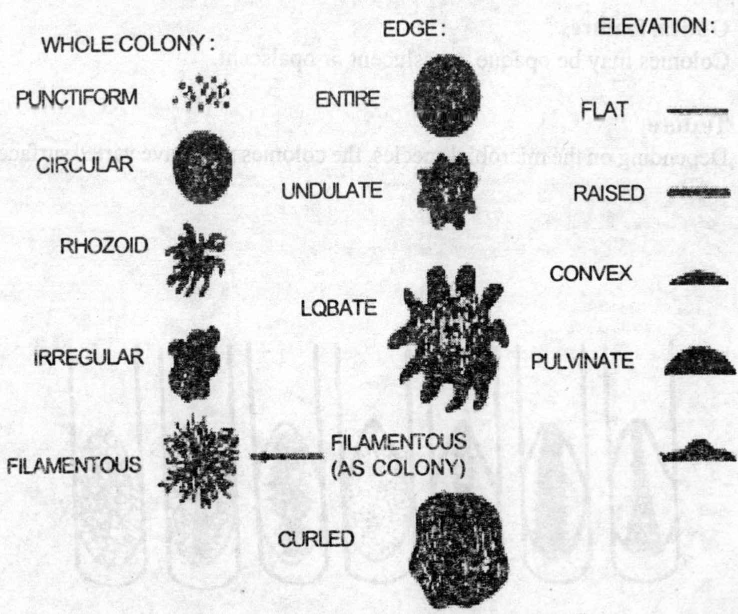

Fig. 1.11 Culture of Microbes
Colony Characters

B. Margin or edge

The contour or the periphery of the colonies may assume various shapes or patterns depending on the microbial species. The margin may have any of the following shapes.

1. *Entire* - Smooth margin
2. *Undulate* - Wavy margin
3. *Lobat* - Lobed margin
4. *Erose*
5. *Filamentous* - Margin extended into thread like growth
6. *Curled* - Folded margin.

C. Elevation

The surface view of the colony is called elevation. The following types of elevations are seen in bacterial colonies.

1. *Flat* - On the same level as the medium
2. *Raised* - Slightly above the level of the medium
3. *Convex* - Raised above in a convex shape
4. *Pulvinate* - Appearing
5. *Umbonat* - Raised above, but with a hood

D. Optical features

Colonies may be opaque, translucent or opalscent.

E. Texture

Depending on the microbial species, the colonies may have varied surfaces like

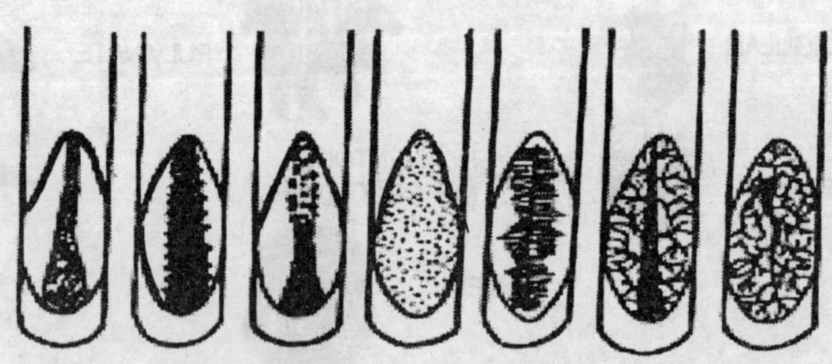

Fig. 1.12 Culture of Microbes
Colony characters (contd) various growth types in streak cultures (for explanation see text)

1. Smooth, shiny or glistening
2. Rough, dull granular or mattle
3. Mucoid (slimy or gummy)
4. Wrinkled or folded surface

It has been found out that for many pathogenic bacteria, the shiny or glistening surface is associated with virulence or toxicity. Colonies which are shiny have capsulated cells, while rough colonies have cells without capsules.

F. Pigmentation or chromogenesis

Many microbes produce intracellular pigments as a result of metabolism and these pigments impart various colouration to the colonies. The following are some of the pigmented colonies (Pelczar, 1986).

Flectobacillus major	Pink
Servatia marcescens	Red
Chromabacterium violaceum	Violet
Staphylococcus aureus	Gold
Micrococcus luteus	Yellow
Derxia gummosa	Brown
Bacteroides melaninogenicus	Black

The pigments produced by bacteria may be *water insoluble and water soluble.* If the pigments are insoluble, only the colonies are coloured, while the medium remains colourless.On the other hand, if the pigments are water soluble, the colour diffuses into the medium and stain it also. For example, in *Pseudomonas aeruginosa,* a blue water soluble pigment (Pyocyanin) stains the medium. Some pigments are partially water soluble (as in *pseudomonas chlororaphis),* and form green crystals (chlororaphin) around the colonies.

Certain types of water soluble pigments are fluorescent i.e., the medium around the colonies show white or green brilliance on exposure to uv light.

Characteristic pigmentation, however, will only develop when all cultural conditions are ideal.

Form : The total appearance of the colony is called form. This is of the following types.

1. Punctiform - Like dots
2. Circular - Like circles
3. Filamentous - With thread like structures constituting the colony
4. Irregular - No definite shape
5. Rhizoid - Thin threads
6. Spindle - Colonies appearing like spindles

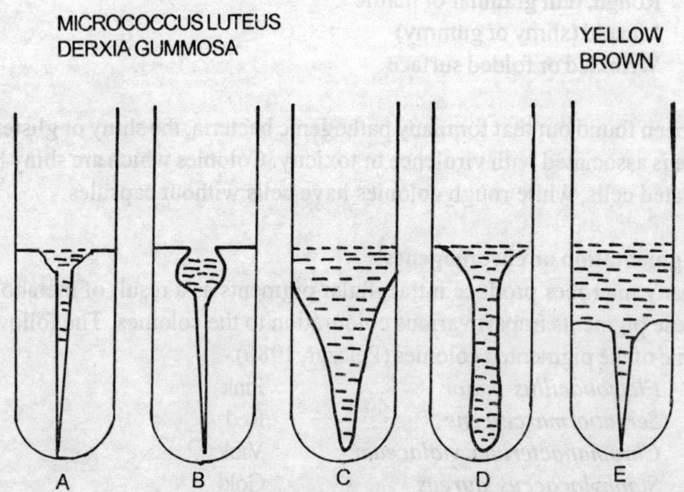

Fig. 1.13 Culture of Microbes

Colony characters (contd) - gelatin stab

Cultures – **A.** Crateriform; **B.** Napiform, **C.** Infundibuliform; **D.** Saccate; **E.** Stratiform

AGAR SLANTS

Cultural characters of bacteria in agar slants are of the following types.

1. Amount - Scanty, moderate or abundant
2. Margin - Entire or irregular
3. Texture - Same as for
4. Form - This is of the following types
   a. Filiform - Like a small tube
   b. Echinulate - With pointed outgrowths
   c. Beaded - Like beads
   d. Effuse - Spread over the slant surface
   e. ·Arborescent - Branched like a tree
   f. Rhiozoid - Thin rihzoid like growth

3. GELATIN STABS

This is basically of two types -

1. **Growth along the line of stab** (puncture) **:** Growth may be confined along the line of inoculation or it may diffuse a little away from it. This is further classified into the following types -
   a. Filiform – Thin tubes
   b. Beaded – Small beads

c. Papillate – Thin finger like projections
d. Villous – Very thin projections
e. Arborescent – Branched like a tree

2. Growth by liquefaction : Here, growth will proceed in various funnel like designs by liquifying the medium along the line of growth. This is of the following types –

a. Crateriform
b. Napiform - Broad above and suddenly tapering below
c. Infundibuliform
d. Saccate -Like a small bag
e. Stratiform

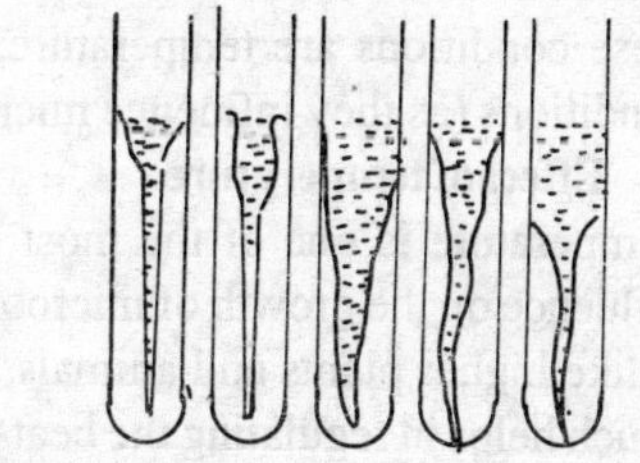

Fig. 1.14 Culture of Microbes
Colony characters (contd) gelatin stab cultures showing line of puncture

3. Nutrient broth : In liquid medium such as nutrient broth, features like shape, form, texture etc. cannot be observed. The growth characters observed are –

a. Amount - Scanty, moderate or abundant
b. Distribution and Type - Cells may be distributed uniformly throughout the medium imparting all round turbidity or it may be confined to the surface or bottom of the broth. The following types of accumulations are seen –

1. Flocculant
2. Ring
3. Pellicle
4. Membraneous
5. Sediment

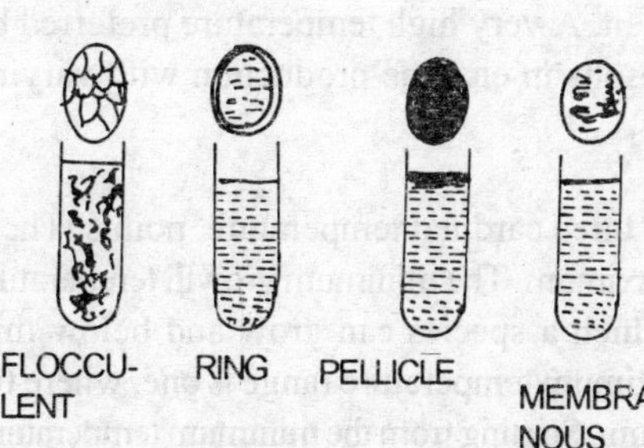

Fig. 1.15 Culture of Microbes
Colony characters (contd) surface growth types in nutrient broth

Some physical conditions required for bacterial growth in a culture media

Apart from the type of media for the bacterial growth, many physical conditions of the environment are very much essential for the optimum growth. Unless these conditions are provided, bacterial colonies do not develop properly. It must, however be pointed out that there is no uniformity for these conditions applicable to all kinds of microbes or all kinds of media. These physical conditions have to be studied for every type of microbe that is to be cultured and the optimum of these conditions has to be fixed for a particular experiment. Some of these conditions are temperature, pH, oxygen, relative humidity etc. These conditions (as they influence microbial growth) are discussed below.

1. Effect of temperature

Temperature is one of the most important physical factors that has a great influence on the growth of microorganisms. Microorganisms like bacteria, are unlike higher plants and animals, in that they lack a homeostatic mechanism which helps in regulating the heat generated by metabolism. Hence, any minor alteration in temperature has a great effect on the growth and physiology of bacteria. Initially, however an increase in temperature range will increase the metabolic rate. But this is, however, subjected to a certain limit. Initially (within a limited temperature range), there is a two fold increase in the rate of enzyme catalysis for every 10^{0}C rise in temperature.

Based on the temperature tolerance and its influence on growth, bacteria may be classified into the following categories.

1. **Psychrophiles** : Optimum temperature ranges between 0^{0}C and 20^{0}C.
2. **Mesophiles** : Optimum temperature between 20^{0}C and 40^{0}C.
3. **Thermophiles** : Optimum temperature between 40^{0}C and 80^{0}C.

Thermophiles are of two categories. These are facultative thermophiles with the optimum temperature range between 45^{0}C to 60^{0}C, and obligate thermophiles with optimum temperature above 60^{0}C. Normally, however, temperature in the range between 50^{0}C and 100^{0}C are lethal for bacterial cells and spores. Some endospores, however can withstand it. A very high temperature preferred by bacteria, is a genetic trait, which results in enzyme production with varying temperature requirements.

For the growth of bacteria there are three cardinal temperature points. These are a minimum, an optimum and a maximum. The minimum growth temperature may be defined as the lowest at which a species can grow and below this, growth comes to a standsstill. An optimum temperature range is one, where the growth of a bacterial species is maximum. Starting from the minimum temperature, an increase to reach the optimum temperature will have a beneficial effect on the growth of bacteria. From the minimum temperature to optimum there is a logarithmic where bacterial growth is still possible but the growth will not be maximum. Any increase beyond maximum will be deleterious. From this discussion it becomes obvious that from the minimum to optimum, bacterial growth reaches the maximum. At the maximum temperature there will be a stand still in the growth rate of bacteria.

For the in vitro growth of bacteria, the duration of exposure at a particular temperature range is very essential. For instance, to decide the lethal effects of high temperature on bacterial growth there are two methods. These are the *thermal death point* (TDP) and *thermal death time* (TDT). The TDP is a temperature at which an organism is killed in 10 minutes of exposure and TDT is the time required to kill cells/ spores suspension at a given temperature.

Effect of pH

The pH is the logermathic symbol of the reciprocal of hyrdrogen ion concentration in gram atmos/per litre. For instance, a pH range of four indicates a concentration of 10^{-4} or 0.0001gm atoms of hydrogen ions in 1 litre of solution (for a detailed discussion of pH and the methods used for its determination – please see College Microbiology Vol. 1, by the same author). In simple terms, a pH may be defined as the relative acidity or alkalinity of a medium or a solution. pH is expressed in a range from 0–14; in this 0 – 7 is acidic and 7–14 is alkaline and 7 itself is called neutral.

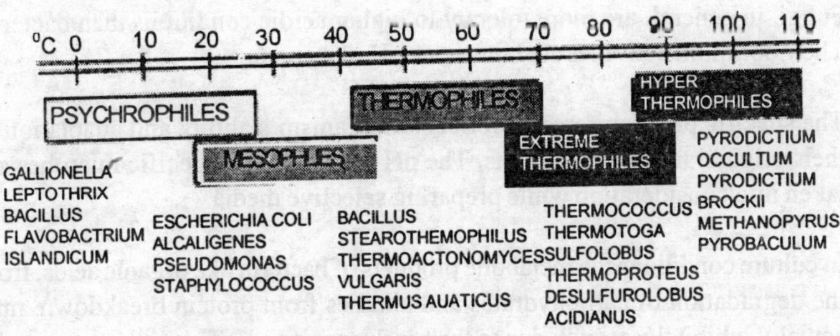

Fig. 1.16 Culture of Microbes
Growth pattern of microbes with reference to temperature

Changes in pH have a profound influence on the growth and metabolism of all organisms including microorganisms. Enzyme systems on which depends all biological reactions are known to be highly sensitive to pH changes. Any change in pH range would render the enzyme non functional, thereby affecting the growth of microbes like the cardinal temperature ranges, the organisms and, the maximum. Above the maximum pH organism fails to grow, and at the optimum pH the growth will be maximum and in the minimum pH growth is just possible. A pH range below the minimum would have a retarding influence on the growth of microbes.

It has to be understood, however, that when the bacteria fail to grow it is primarily due to the non functioning of the enzyme in a pH range that is unsuitable to them.

There is no general, minimum, optimum and maximum pH range that can be fixed as standard for the growth of microorganisms. Different microbes have different pH ranges suitable for them. And each species has a preference for a specific pH range that may be broad or limited. Most of the times optimum growth of a microbe occurs within a narrow optimum pH range. Experiments have shown that for many bacteria, the specific pH range is between 4 and 9. With the optimum ranging between 6.5 - 7.5 pH. There are, however, a good number of acidic pH. For instance, *Mycoderma aceti* and *Thiobacillus thioxidans* prefer an acidic pH below 4. *T. thioxidans,* infact prefer a full acidic pH ie., 0. This is the reason why organic acids are used to preserve foods like fermented dairy products.

Fungi, in general, are more tolerant to higher acidic conditions than bacteria. Their optimum activities are between a pH of 4 and 6.

The specific pH requirements reflect an organism's ability and adaptation to their original natural conditions. The pH preference of specific microbes are taken into consideration while preparing selective media.

In culture conditions the metabolic products of bacteria like organic acids, from the degradation of carbohydrates and alkalies from protein breakdown, may actually inhibit the growth due to their influence on pH. This is the reason why buffers are added to the media to take care of acidic or alkalinity influence and to neutralise them. Generally amphoteric compounds such as, peptones and aminoacids found in the complex media act as natural buffers. In chemically defined media phosphates salts are often used to function as buffers.

Effect of relative humidity

Water is one of the basic ingredients necessary for the growth of all microorganisms. Some microorganisms however, are less sensitive to relative humidity (this may be defined as the amount of water vapour present in the air expressed as a percentage of maximum that the air could hold at a given temperature) and can survive in the absence of required quantity of moisture.

Such microbes, however, recommence growth from hybernation at the availability of suitable moisture conditions. There are some microbes which are extremely sensitive to moisture and die in the absence of it. The different water requirements of the microorganisms can be studied in the the laboratory under different moisture conditions.

Different relative humidity conditions can be created in the laboratory by using pure water vapours, or water saturated with certain chemicals where in the relative humidity comes down. For instance, pure water vapours will create 100% relative humidity (RH). Chemical solution containing $CaSO_4\ 5H_2O$ gives 95% RH. Growth of microorganisms get generally inhibited when the RH is below 62%.

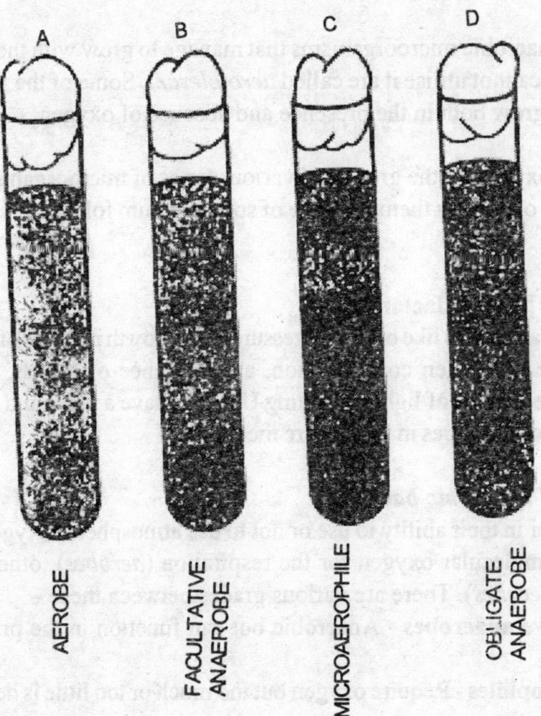

Fig. 1.17 Culture of Microbes

Effect of oxygen

Based on their requirement of oxygen, microorganisms can be classified into three major groups – these are *aerobic, anaerobic* and *microaerophilic.* Microbes which require the presence of oxygen for their growth are called aerobic; those which can grow only in the absence of oxygen are called anaerobic, and those that require limited amount of oxygen and increased amount of carbon dioxide (7-10%) for their growth, are known as microaerophilic. Aerobic and anaerobic microorganisms include obligate and facultative individuals. Obligate aerobes such as, acetic acid bacteria can grow only in the presence of molecular oxygen as they respire aerobically. In contrast with this, there are obligate anaerobics like *Clostridium,* which can grow only in the absence of oxygen because they lack catalase and the resultant accumulation of hydrogen peroxide H_2O_2 is lethal to them.

Some of the anaerobic microorganisms that manage to grow with the presence of oxygen but cannot utilise it are called *aerotolerant.* Some of the facultative microbes can grow both in the presence and absence of oxygen.

The effect of oxygen on the growth of various types of microorganism can be determined by observing them in a tube of solid medium following shake tube inoculation.

Effect of other physical factors

Several other parameters like osmotic pressure, salt, growth patterns of microbes with reference to oxygen concentration, and presence of metals, dyes and different wave lengths of light, including UV rays have a profound influence on the growth of microbes in the culture medium.

Cultivation of anaerobic bacteria

Microbes differ in their ability to use or not to use atmospheric oxygen. While some require molecular oxygen for the respiration (*aerobes*). others do not require it (*anaerobes*). There are various grades between these –

1. **Facultative anaerobes** - Anaerobic but can function in the presence of oxygen.
2. **Microaerophiles** - Require oxygen but too much or too little is detrimental to them.
3. **Obligate anaerobes** - Strictly anaerobic, may even die in the presence of oxygen.
4. **Aerotolerant** - Anaerobic, survive in air, but do not reproduce.

Thus one can see that a wide array of bacteria can be categorized based on their reaction to the presence or absence of oxygen. This is due to the different enzyme systems present in these types of bacteria.

Several methods are employed by microbiologists to selectively grow anaerobic bacteria. Some of thesc are described below.

1. Pre reduced medium

This involves freeing the culture medium of oxygen prior to inoculation (removal of oxygen is said to be reduction). During preparation, the culture medium is boiled for a few minutes to drive out most of the dissolved oxygen. Reducing agents such as, cysteine, is added to the medium to further decrease the level of oxygen content. Oxygen free N_2 is bubbled through the medium to maintain anaerobic conditions. The medium is then distributed to suitable culture tubes which are also flushed with oxygen free N_2. The tubes are further autoclaved and stored before use. At the time of inoculation, the tubes are flushed with oxygen free CO_2 by means of a cannula; plugged again and incubated.

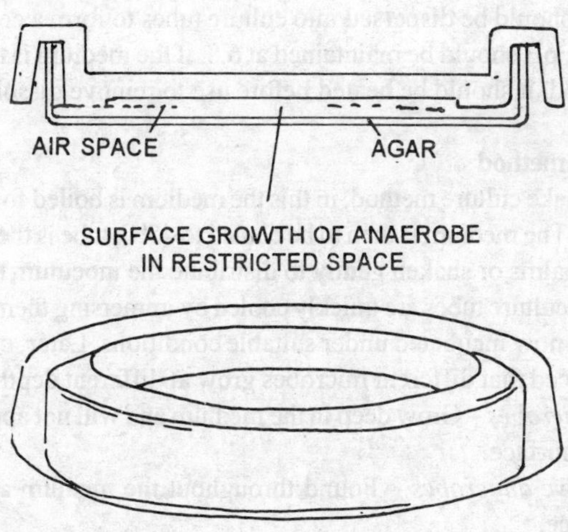

Fig. 1.18 Culture of Microbes
Anaerobic culture plate (air trapping technique)

2. Air trapping method

Air trapping method is one of the most effective method of absorbing oxygen from a layer of trapped air by the use of a special hood or cover. This special cover when kept on a specially prepared agar medium, covers the medium all over except for a thin gap over the medium where air gets trapped. Special chemicals will be added to the medium which will absorb the trapped air (oxygen). Some of these chemicals are – sodium thioglycollate (0.1%) or cysteine, and hydrochloride (0.2%). By this procedure oxygen content of the medium is considerably reduced and only anaerobic microorganisms grow. A special anaerobic medium given below is normally used for culture.

Thioglycollate dextrose medium

Yeast extract	5.0g
Casitone or Trypticase	15.0g
Sodium chloride	2.5g
1-cysteine	0.25g
Sodium thioglycollate	0.5g
Agar	0.75g
Methylene blue 2%	0.1g
Distilled water	1 litre

The medium should be dispersed into culture tubes to form a column about 3-4 inches deep, pH should be maintained at 6.8, if the medium is to be stored for a longer period, it should be heated before use to remove dissolved air.

3. Roll tube method

Also called shake culture method, in this the medium is boiled for 10 minutes to expell the air. The medium is then to be cooled to 45°C, tube is then gently rolled between the palms or shaken gently to distribute the inoculum throughout the medium. The culture tubes are quickly cooled by immersing them in cool water. The tubes are now incubated under suitable conditions. Later, on observation, it will be noticed that different microbes grow at different depths as follows –

1. *Strict anaerobes* – Grow deep in the medium and will not appear anywhere near the surface.
2. *Facultative anaerobes* – Found throughout the medium as well as near the surface.
3. *Microaerophilic* - Grow a little below the surface.
4. *Aerobes* – Grows on the surface of the medium.

The medium used in the above experiment is tryptone yeast glucose agar medium.

4. Deep media method

In this method, aerobic tissues or cells such as germinating seeds, plant cells, fruit or other oxygen using material is added to the medium (broth). After sometime, the medium is boiled and cooled to 45°C and inoculated with a suitable anaerobic microbe. Strict anaerobic microorganisms develop deep in the medium.

5. Phosphorous stick method

Petriplates containing culture media are kept in a closed air tight vessel. Phosphorous sticks are burnt in the vessel to remove oxygen. The sticks continue to burn till all the oxygen gets exhausted and ultimately only the inert gases and partial vacuum remain. A small container of water should be kept to trap P_2O_3, formed as a result of combustion.

6. Pyrogallol method

Certain chemicals such as, a mixture of NaOH and pyrogallol are known to absorb oxygen in a closed vessel where petridishes can be kept. Pyrogallic acid reacts with oxygen and leaves only inert gases, nitrogen and partial vacuum. The procedure is as follows :

(a) Take two agar slants and lable them with the microorganism to be inoculated.
(b) Inoculate the medium under aseptic conditions.
(c) Replace the cotton plug and ignite it with the flame of a bunsen burner.
(d) Using a glass rod, push the burning cotton plug deep into the tube until it nearly touches the surface of the slant.
(e) Put crystals of pyrogallol into the tube on the cotton plug. Add 2ml of sodium hydroxide (4%) into the tube.
(g) Invert the tubes and incubate them at the inverted position at 37°C for 24–48 hours.
(h) Growth of anaerobic bacteria may be observed on the slants.

7. Anaerobic chamber method

Oxygen from the medium may be completely removed by using special anaerobic jars (chambers), such as Brewer's jar, Brown's jar. McIntosh jar, Fields jar or Gas pack system. In these H_2 CO_2 or N_2 can be introduced through a special inlet. The jars have a special air lock where in the culture plates are kept. The air locks are evacuated and refilled with N_2. From the air lock, the plates are kept into the main chamber. Hydrogen is slowly introduced into the medium. This reacts with whatever O_2 is present, forming water. This reaction is catalysed by heating electrically a palladium catalyst. Water, that is formed in this reaction is absorbed by $CaCl_2$. After this, the medium is inoculated with a suitable anaerobic bacteria.

Another device that is simpler than the one explained above is a gas pack system. The gas pack consists of dry chemicals which when hydrated release CO_2 and H_2.

A gas pack as the one mentioned above, is introduced into a sealed jar which contains an inoculated medium. Addition of water (10ml) into the gas pack causes the release of CO_2, and CO_2. H_2 reacts with O_2 on the heated palladium catalyst to form water bringing in anaerobic conditions. The CO_2 evolved also aids in the growth of strict anaerobics. An anaerobic indicator ship introduced into the system changes from blue to colourless in the absence of oxygen.

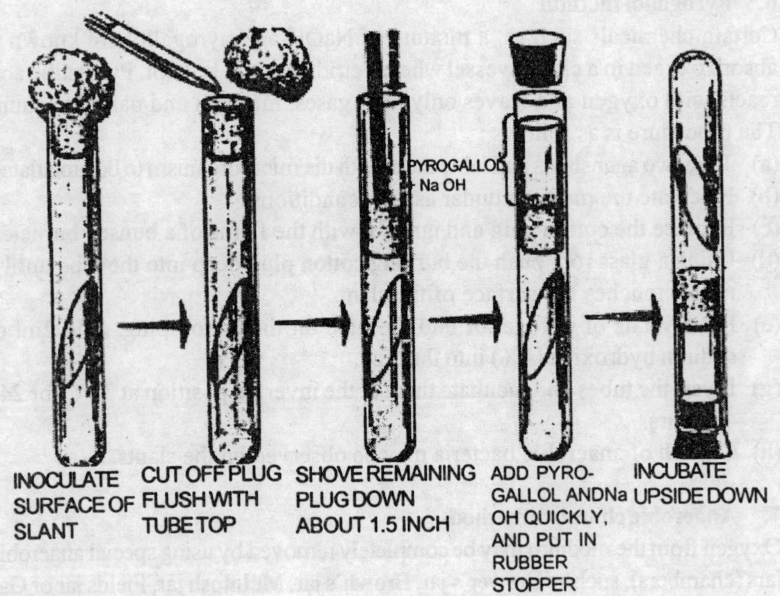

Fig. 1.19 Culture of Microbes

Pyrogallol NaoH method for anaerobic cultivation of bacteria

ISOLATION OF SOIL ALGAE

Thousands of types of algae occur in nature in all possible types of environmental conditions. They abound however, in oceans, salt lakes, freshwater lakes, ponds and rivers. Some algae are also found on damp soils, rocks, tree barks and even in the bodies of plants and animals. In size, algae range from extremely minute to the largest. Some of the sea kelps are more than 90 feet in length. Algae are both prokaryotic (if blue green algae are included), and eukaryotic.

In a microbiological laboratory for the purpose of study, pure cultures of algae are necessary. In general, techniques used for bacteria and fungi are applicable to microalgae. The medium normally used for isolation of soil algae is Beneck's broth. The following are the requirements to prepare Beneck's broth medium.

KNO	0.2g
$CaCO_3$	0.1g
$FeCl_3$	2.0 drops
Glass distilled water	1000.0 ml

Suitable soil sample

Procedure : Required amounts of the chemicals mentioned above are dissolved in 200ml of distilled water, and the volume is made up to a litre by the addition of more distilled water. The above components will give Beneck's liquid medium. Solid medium can also be prepared by the addition of 50 x 20gms of agar per litre of Beneck's solution. The pH of the medium should be adjusted to between 7 and 7.5.

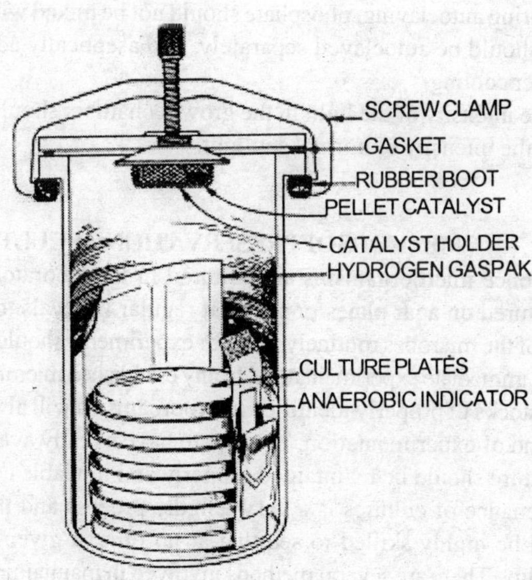

Fig. 1.20 Culture of Microbes
The gas pack system for cultivation of anaerobic bacteria

Algal cultures may be prepared as follows :

1. Pour the broth solution into suitable conical flasks. The conical flasks are filled to a depth of 5cms from the bottom. Plug the mouth of the flasks with cotton wool.
2. Autoclave the medium in conical flasks at a pressure of 20lb for 20 minutes.
3. Allow the medium to cool and remove them from autoclave.
4. Remove the cotton plug from the mouth of conical flasks and add one gram each of soil; immediately replace the cotton plugs. These operations have to be carried out very quickly and under aseptic conditions to prevent contamination.
5. Incubate the culture medium with the inoculum at 30-35^{0}C in an illuminated growth chamber (usually fitted with 60w tungsten lamps).

Observation : After about 2 weeks of incubation, profuse growth of algae can be seen on the medium.

The algal cultures obtained from soil are mixed cultures. They may be further purified by carefully transfering a few cells of filaments under aseptic conditions to a different culture medium. These can be further subcultured.

1. During autoclaving, phosphate should not be mixed with other ingredients. It should be autoclaved separately, and aseptically added to the medium after cooling.
2. The intensity of the light in the growth chamber should ordinarily be 1/3rd of the intensity of normal sunlight.

METHODS OF PRESERVATION OF CULTURES

When once microorganisms are cultured in the laboratory, they are usually subcultured on agar plates or slants at regular intervals to maintain viability. Some of the microbes routinely used for experiments should always be available for any immediate experimentation. In any established microbiological laboratory, ready stocks of properly identified microorganisms will always be available for any kind of experimentation. In order to have a ready availability of cultures, the culture should be maintained properly and in viable condition.

Maintenance of cultures is a very lengthy process and the persons involved should be highly skilled to see that at no time, a given culture should lose vialibility. There are several methods involved in maintaining the cultures. Some of these are described below.

1. Use of refrigator or cold room storage

Live cultures on a culture medium can be successfully stored in refrigators or cold rooms, when the temperature is maintained at 4^{0}C. At this temperature range the metabolic activities of microbes slows down greatly but do not all together stop. As a result, bacterial metabolism will be very slow and only less quantity of nutrients will be utilized. This method, however cannot be used for a very long time for not only the nutrients get utilized but also waste products get accumulated killing the microbes. Refrigerator or cold room storage is of use only for short time preservation of cultures. Subculturing is necessary if the period exceeds two weeks.

2. Paraffin method

This is a very simple and cost effective method of preserving cultures of bacteria and fungi for several years at room temperature. In this method, sterile liquid paraffin is poured over the slant culture of microbes, and stored upright at room temperature. The layer of paraffin prevents dehydration of the medium and ensures anaerobic conditions. As a result, the microbes remain in dormat condition. It has been seen that in some instances bacteria remained viable for even upto 15 or 20 years. Cultures can also be maintained by covering the agar slants with a layer of sterile mineral oil about 1/2inch above the surface of the slant.

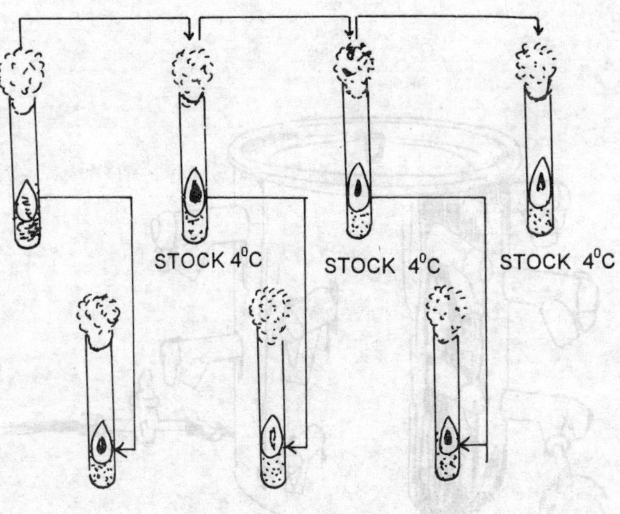

Fig. 1.21 Culture of Microbes
Preservation of culture by periodic transfer

Maintenance by periodic transfer of culture

The usual method of maintaining a culture is to inoculate the microbe in agar slant and incubate the medium for a short period. This gives sufficient time for the microbes to reach the optimum growth phase. Subsequently, the culture can be stored at a low temperature. As and when required, they can be taken out and cultured on a fresh medium and then stored again under cold temperature conditions. Periodic subculturing, however, might induce mutations in the microbes.

Preservation at - 40°C in glycerol

Cultures can be preserved for a number of years in glycerol, at a temperature of - 40°C in a freezer. In this method, about 2ml of glycerol solution is added on to the agar slant culture. The culture can be emulsified by shaking the culture. Emulsion is then transfered to ampoules, with each ampoule having 5ml of the culture. These ampoules are placed in a mixture of industrial mathylated spirit and carbon dioxide and are frozen rapidly to - 70°C. The ampoules are then removed and placed directly in a deep freeze at - 40°C. For utilization of the stock cultures, the ampoules are kept in a water bath at 45°C for about a few seconds and then used for plate cultures.

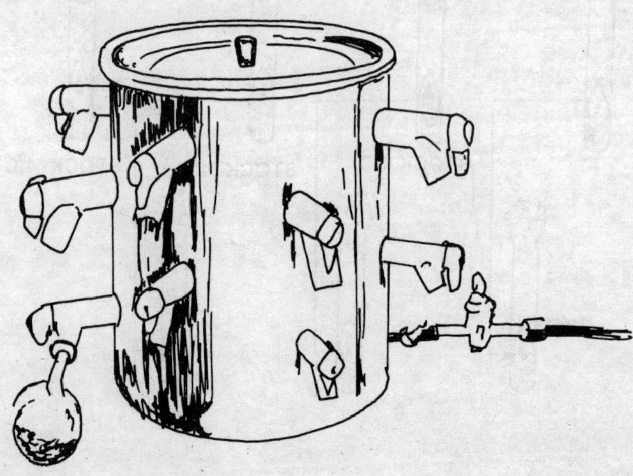

Fig. 1.22 Culture of Microbes
Preservation of cultures by lyophilization

Freeze drying

This method is also called lyophilization. Lyophilization, may be defined as the rapid dehydration of the organisms while they are in a frozen stage. In this method, a thin film of frozen bacteria are dried in a vacuum and subsequently sealed in ampoules. Freeze drying is one of the most frequently used methods of culture preservation employed in many microbiological laboratories all over the world.

Lyophilization protects the damage caused to microbial cells during the process of dehydration. As the organism is freeze dried, the metabolic rate greatly slows down and the microbes remain in a dormant stage for several years.

Generally in the process of drying, elimination of water results in damage to the many chemical compounds and the integrity of the cell is lost. In the freeze drying process, however, the culture is frozen first and water is removed by sublimation from the ice as vapours. As the suspension is not in the liquid condition, the process does not damage the structure. The freeze dried materials can easily be brought back to the normal stage without any damage to the original condition.

In the freeze drying process, the cultures are rapidly frozen at - 70°C, and then dehydrated by vacuum. The tubes are then sealed and stored in the dark in refrigerators at 40°C.

Freezing in liquid nitrogen at temperatures of - 196°C also retards the metabolic rate and the cells remain viable for very long periods.

Freeze drying involves several steps. These are the following –

Predrying requirement for cultures

The type of culture media used is an important criterion in the freeze drying cultures. For some microbes, the predrying culture and maintenance may be the same or different. For example - for *Bacillus macroides,* the predrying culture and maintenance culture are the same, whereas for *Bacillus subtills,* the predrying culture is soil extract agar whereas for the maintenance it is nutrient agar.

The predrying media should have a rich concentration of the microbes. The age of the culture is another important criterion because the cultures that have reached the optimum growth phase survive better than the cultures that are still in the growth phase.

Ampoule preparation

The ampoules used in the preservation of the culture should be made of neutral glass and have an internal diameter of 6mm. Ampoules should first be disinfected and then rinsed in distilled water. After plugging with the cotton wool, they should be sterilized with an autoclave for 20 minutes under 20 pounds pressure.

Harvesting the culture

Cultures grown on agar slants should be harvested for 3-5 days after incubation. 2ml of suspending fluid such as, *Mist Desiccans* containing sterile horse serum, glucose and nutrient broth should be added over each agar slant and the surface should be gently rubbed with a Pasteur's pipette. The suspensions should immediately be transfered to the ampoules.

Primary drying

The filled ampoules should be placed in a stand and the neck of the tube should be flamed. The ampoules should then be centrifuged. Special centrifuges are used for this purpose. These centrifuges have a primary drying chamber and come in a variety of models. Most of these models consist of a drying chamber, where the centrifuge head is located and heated under vacuum. The centrifuge head should be allowed to run at 500rpm. Usually, the primary drying is allowed to proceed between $2^{1/2}$ to 4hrs during which, more than 90% of free water is removed. Air is then allowed to slowly enter into the vacuum chamber. The centrifuge head is then removed from the machine and the ampoules are plugged again with cotton wool.

Secondary drying

The ampoules are then transferred to a secondary dryer which consists of high grade phosphorous pentoxide contained in a tray, to which is attached a vacuum pump. The ampoules are left on this dryer for 18-20hrs upon which the moisture content is reduced to 1%. The ampoules are checked for maintenance of vacuum, and sealed with flame. The culture in the ampoule is now a light powdery substance. The ampoule can be stored at 4°C.

Reconstitution of the culture from ampoules

At the time of utilization, the tip of the ampoule may be cut using a file. The cotton plug should then be carefully removed with the help of forceps. The mouth of the ampoule should be flamed again and a fresh sterilised cotton wool can be plugged.

Sordelli's method of preservation of cultures

This is a method of preservation simpler than the freeze drying method, but as reliable as the latter. This method can be safely used when the samples to be preserved are small in quantity.

In this method, the culture should be preserved or incubated on a solid medium for the required period. The inoculum is emulsified in a loopful of horse serum and is deposited on the inner wall of the small tube (8 x 60mm) which is then put into another larger tube (10 x 150mm). A small quantity of phosphorous pentoxide is placed at the bottom of the outer tube with the help of a glass rod. The inner tube must be placed in such a way that it is held over the bottom of the outer tube but not directly touching the chemical placed at the bottom. The outer tube, is then connected to a vacuum pump, and after the air is removed, the outer tube is sealed. This tube containing the culture can be stored at room temperature away from light.

Preservation of soil cultures

Culture of soil microbes can be maintained in the soil medium. The method of preservation is as follows. A sample of soil is passed through a sieve of 2mm mesh, and collected in sufficiently large test tubes. To these tubes is added 1% solution of dextrose and mixed thoroughly. These tubes are then kept in a boiling water for about 15 minutes and then autoclaved for about one hour at 13 pounds pressure on two successive days.

Pure culture of soil microbes can be kept in this medium and preserved for a number of months under refrigeration.

Preservation of plate cultures

The following are the various methods for the preservation of cultures on petriplates.

1. Put a drop of formaldehyde on the inside of the lid. The formaldehyde solution acts as fixative and will not alter the growth of the already existing colonies.
2. The plates can also be kept under refrigerator at 4°C which will retard further growth.
3. The petriplates temporarily sealed with vaseline which will prevent dehydration of the medium. After the atmosphere inside the media gets exhausted; colonies will not grow further.
4. The clean embedding resin may be poured over the surface of the plate culture this will also help in preservation.

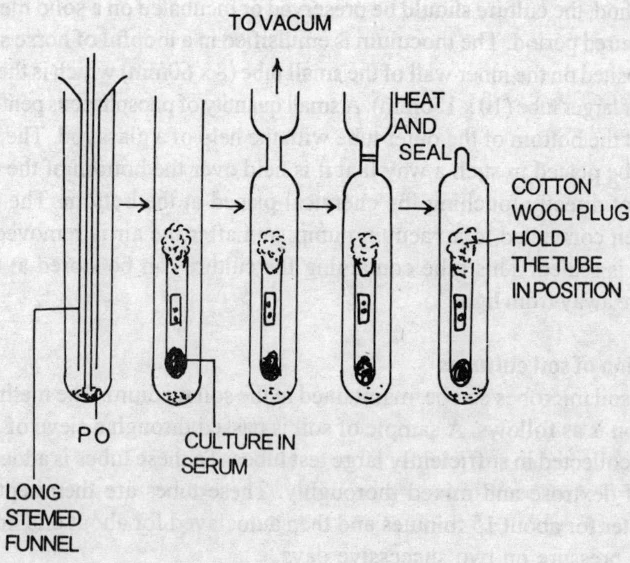

Fig. 1.23 Culture of Microbes

Preservation of cultures by Sordelli's method

Preservation of fungal cultures of fumigation

In fungal cultures infestation by mites is a serious problem of maintenance. Mites get into the culture through various means, such as soil, plant materials etc. The mites that develop on the culture feed on the fungal hyphae and move from one culture to the other through cotton plugs. The development of mites can be prevented by addition of 0.01% lindane powder (an insecticide) to the sabouraud dextrose sugar medium. This kills the mites within 3 minutes. Alternatively the cultures may be dry fumigated with pra-dichloro-benzene

2

MICROBIAL GROWTH AND REPRODUCTION

One of the unique properties of living organisms is their ability to grow and reproduce. It is this unique capacity that has been responsible for the maintenance of life ever since it originated on this planet a few billion years ago.

Microorganism growth is simply an increase in size as most of them are made up of only one cell In unicellular microorganisms cell division is a means of multiplication (reproduction), and growth. Whenever a cell division occurs, the two daughter cells become two individuals; the cells, however, increase their size (grow) and attain maturity.

Nutritional requirements

Microorganisms, in order to grow must obtain from the environment, all the substances that they require for the synthesis of the cell materials and for the generation of energy. These substances are termed nutrients. In nature, microbes draw from their habitats all the necessary *nutrients*. While culturing bacteria, however, these nutrients must be made available to them in the medium. In addition to nutrients, water plays a key role in the culture of microbes. Although water is not regarded as a nutrient by itself, its presence in the medium is a must, as otherwise biological reactions will never take place. A microbe can absorb nutrients only in the form of a solution.

Microbes are extremely diverse in their nutritional requirements, not only in terms of quality, but in terms of quantity also. A culture medium must, therefore have all the required materials for the growth of a particular microbe. Practically thousands of media have been proposed for the cultivation of bacteria, often with reasons for the presence of a particular nutrient.

The design of a culture medium, therefore should be based on *Principles of nutrition* and must fulfill the minimal nutritional requirements that are essential for the synthesis of cell substances.

Elementary nutrient requirements

The chemical composition of cells, which is broadly constant throughout the living world indicates the materials required for microbial growth. Water accounts for some 80 to 90 per cent of the total weight of cells, and is a major nutrient in quantitative terms. Besides water, the elementary composition of the cells is divided into ten macro elements which are present in all cells. These are carbon, hydrogen, oxygen, nitrogen, sulphur, phosphorous, sodium, potassium, calcium, magnesium and iron (C, H, O, N, S, P, K, Na, Ca, Mg, Fe). Of these, hydrogen, oxygen, carbon, nitrogen, phosphorous and sulphur account for 95 per cent of the cellular dry weight. Trace elements required for microbial growth are – manganese, molybdinum, zinc, copper, cobalt, nickel, vanadium, boron, chlorine, selenium, silicon, tungsten etc. Heavy metals are mostly part of the enzyme systems, involved in the metabolism of inorganic elements and compounds.

All the required metallic elements can be supplied as nutrients in the form of cations of inorganic salts. Potassium, magnesium, calcium and iron are required in relatively large amounts and should be included as salts in the media. Phosphorous, a non metallic element can be provided as phosphate salt.

It should be understood, however, that some biological groups have additional specific mineral requirements; for example, certain marine bacteria require high concentration of sodium.

Carbon requirement

Photosynthetic and chemosynthetic bacteria typically use CO_2 as the principal source of their carbon requirement. The conversion of inorganic carbon into organic cellular constituent is a reductive process requiring the input of energy. In phototrophic and chemotrophic bacteria, the energy obtained from light or by the oxidation of inorganic compounds is used to reduce CO_2 to carbohydrates. Heterotrophic bacteria, however, obtain their carbon requirement directly from organic compounds. In this process, however, there is no need for any further reduction, as all organic substrates are at the same general oxidation level like the cell constituents.

The carbon that enters the cell system has to meet two requirements of the cell.

1. Carbon needed for the biosynthetic pathways of the cell.
2. Provide the energy necessary for the biological activities of the cell. For

this, the carbon compounds directly enter into energy releasing metabolic pathways ultimately releasing CO_2.

In the case of aerobic bacteria, organic carbon gets completely metabolized and CO_2 is released, in the case of fermentive bacteria, however, as the organic carbon is incompletely oxidized, a mixture of CO_2 and organic compounds are excreted.

Many of the microorganisms, such as *Escherichia coli* can use a single carbon source (organic compound) as the starting point, and using their synthetic machinery can produce all their requirements. Some other microbes, however, require more than one carbon source as their primary requirement. This is due to the fact that their metabolic machinery cannot synthesize all their diverse carbon requirement. As a result, some of the required ones have to be added directly to the medium. These additional organic nutrients have a purely biosynthetic functon being required as precursors of certain organic cell constituents. These are often called the growth factors.

Microorganisms are extremely diverse, both with respect to the kind and number of organic compounds which they can use as the main carbon source. Some microorganisms are highly versatile in that they can use any of the carbon compounds as their energy source, while some are highly specialized and require a specific organic compound for their growth. For example, certain cellulose decomposing bacteria can use only cellulose.

Some microbes require gaseous CO_2 itself as a nutrient in very small amounts. This requirement however, is met by the CO_2 naturally released by the organisms themselves during metabolism. It has been known that a few bacteria and fungi require a relatively high concentration of CO_2 (5–10 percent), for growth in the media.

Nitrogen, Sulphur and Phosphorous requirements

Nitrogen and Suphur occur in the organic compounds of cell, mainly as amino group or sulphydryl groups in the aminoacids (proteins). Most of the photosynthetic and many of the non photosynthesizing bacteria take up these elements (Nitrogen and Sulphur) in the oxidized inorganic state as nitrates and sulphates; thus their biosynthetic utilization requires a preliminary reduction. There are certain microbes which cannot bring about a reduction before uptake, and they have to be provided N_2 in a reduced form (ammonium salts). Reduced sulphur requirement can be met by inorganic compounds such as sulphides.

Nitrogen and Sulphur requirements can be met by organic compounds that contain these elements in the reduced state. Amino acids and other protein degradation products (Peptones), are the source of organic nitrogen and sulphur. Some of these compounds can also provide simultaneously the carbon requirements of the bacteria.

There are some unique bacteria which can use gaseous nitrogen of the atmosphere as their nitrogen source (Eg., *Clostridium, Azatobacter, Rhizobium* etc). Such bacteria are termed *Nitrogen fixers* and the process involves a preliminary reduction of N_2 to ammonia. Some of the cyanobacteria (*Nostoc, Anabena* etc) are also capable of biological nitrogen fixation.

Within the cell, phosphorous occurs as phosphate and phosphate esters. Phosphates occurs in the phospholipids of membranes and phosphate diesters from the backbone linkage of micromolecules such as nuclei acids. The role of phosphates needs hardly to be emphasized. They participate in all energy conversion reactions.

Other mineral requirements

In addition to the four major elements (C, N, P and S), microorganisms also need a large number of minerals in trace quantities. Most of these trace elements are required for the functioning of the enzyme systems. Each of these minerals perform a specific function and their presence in the medium in optimum concentration decides the growth of microorganisms. The following table gives a summary of the role of some minerals in microbial nutrition.

Some physiological functions of trace elements

Elements	**Functions**
Potassium	Cofactor of some enzymes
Magnesium	Cofactor in enzymatic reactions, binds enzymes to substrates, constituent of chlorophyll
Iron	Constituent of cytochromes and other heme and non-heme proteins, cofactor of many enzymes
Calcium	Cofactor for enzymes such as proteinases
Magnesium	Cofactor for some enzymes, can replace Mg in some cases
Copper, Zinc, Molybdenum	Inorganic constituents of some enzymes
Cobalt	Inorganic constituents of some enzymes

Role of Oxygen : Aerobic bacteria and other microorganis require molecular oxygen for their respiration (for details see chapter 1).

Growth factors

Organic nutrients, that cannot be synthesized by the microbe, but essential for its growth are collectively called growth factors. Most of the growth factors fall into one of the following three classes (Stainer et al, 1986) –

1. *Amino acids* – Required as constituents of proteins.
2. *Purines and pyrimidines* – they are neccssary as constituents of nuclei acids.
3. *Vitamins* – These present a diverse group of organic compounds constituting the prosthetic group of some enzyme systems. The following is a table of vitamins and their role in microbial metabolism.

Table

Relation of some water -soluble vitamins to Coenzymes

Vitamin	**Coenzyme**	**Enzymatic reactions involving the Coenzyme Form**
Nicotinic acid (Niacin)	Pyridine nucleotide coenzymes (NAD + and NADP +)	Dehydrogenations
Riboflavin (Vitamin B_2)	Flavln nuclcotidoo (FAD and FMN)	Some dehydrogenations, electron transport
Thiamin (Vitamin B_1)	Thiamin Pyrophiosphate (Cocarboxylase)	Decarboxylations and some group-gransfer reactions
Pyridoxine	Pyridoxal Phosphate	Amino acid metabolism Transmission Deamination Decarboxylation
Pantothenic acid	Coenzyme A	Keto-acid oxidation, fatty acid metabolism
Folic acid	Tetrahydrofolic acid	Transfer of one-carbon units
Biotin	Biotin	Co_2fixation, carboxyl transfer
Cobalamin (Vitamin B_{12})	Various cobalamin derivative	Molecular rearrangement reactions

Growth factors are required by microorganisms in extremely minute quantities as they satisfy some specific requirements.

Nutritional forms

Microorganisms obtain nutrition by different methods. This constitutes the basis for their categorization into various types depending on the source of obtaining their nutrition. With reference to obtaining inorganic nutrients, all microorganisms obtain them directly from the environment in their finished form. However, with reference to the organic metabolites, there are various patterns by which microbes procure them from the environment. Basically from the point of view of the source of nutrition whether inorganic or organic, microorganisms can be divided two categories – *autotrophs* and *heterotrophs*. Autotrophic microorganisms can survive in an exclusively inorganic environment. Using this as the source of basic energy, their cellular machinery can synthesise all their organic nutritional requirements. Heterotrophic organisms on the other hand require an input of readymade organic food as they do not possess the synthetic machinery to produce organic metabolites by the reduction of carbon.

Autotrophic microorganisms, which can manufacture food from inorganic sources require two basic ingredients – 1. A supply of inorganic raw materials 2. A source of external energy to fabricate the inorganic materials into organic nutrients. Based on the mode of the input of the external energy, autotrophic microorganisms can further be classified into *photosynthesizers* (photoautotrophic), and *chemosynthesizers*. Photosynthesizing microorganisms utilise solar energy to reduce the inorganic carbon into organic carbon. Chemosynthesizing microorganisms on the other hand utilise the oxidation of inorganic rawmaterials as a source of external energy for the reduction of inorganic carbon.

Microorganisms can be classified either based on the method of food procurement (Autotrophic or heterotrophic), or they may be also classified from the point of view of the source of input of external energy (photosynthetic or chemosynthetic). A categorisation of microorganisms is also possible taking both the bases mentioned above as a criteria for classification. On this basis, the following categories of microorganisms can be identified.

AUTOTROPHIC MICROORGANISMS

All types of phototrophic microorganisms use the environmental carbondioxide as source of their inorganic carbon. This is reduced to carbonhydrates by the addition of hydrogen obtained from water or other sources. Autotrophic microorganisms are classified into following categories.

1. Photoautotrophs

Most of the pigmented bacteria and microalgae belong to this group. In all the photosynthesizing microbes, solar radiation is the basic energy source and to trap this radiation, one are more varieties of chlorophyll are present in the organisms.

In all photoautotrophic microbes, the source of hydrogen is water except in photoautotrophic bacteria.

In the case of photoautotrophic bacteria, the source of hydrogen is not water and as such oxygen is not a byproduct of photosynthesis. Bacteria such as these usually live in sulphur rich environments (sulphur springs), where hydrogen sulphide is normally in abundance. This substance serves as the hydrogen source for photoautotrophic bacteria.

Two groups of photoautotrophic bacteria have been identified, based on the type of pigmentation and other features. These are the **purple sulphur bacteria** and **green sulphur bacteria.** The purple sulfur bacteria possess a special type of chlorophyll known as *bacteriochlorophyll,* which is green colour but the colour is masked by yellow carotenoid pigments. Green sulphur bacteria possess the pigment *chlorobian* or *bacteriviridin.* This pigment is different from bacteriochlorophyll. Additionally, the green colour here is not masked by carotenoids.

Light ——— Special chlorophyll ——— Energy

H_2S ——— S (byproduct)

H_2

CO_2 ——— Carbohydrate

In the above mechanism of photosynthesis, elemental sulphur is the byproduct. It gets stored inside the cells in purple sulphur bacteria, and in the case of green sulphur bacteria it is excreted.

2. Chemoautotrophs

Chemoautotrophs microorganisms are entirely made up of bacteria. Generally, they are pigmentless and as such there is no mechanism in them to use the light as a source of energy. In addition to the input of inorganic carbon, they have a variety of inorganic sources whose oxidation provides the necessary energy. These inorganic metabolites, in most of the instances are combined with oxygen in the cells resulting in the release of energy and a variety of inorganic byproducts. In the subsequent food manufacture, water and carbon dioxide act as the inorganic raw materials. The general pattern of food manufacture in chemoautotrophs is as follows :

+0

Inorganic Metabolites ——— Inorganic Byproducts

Energy

O (byproduct)

H_2O —————————— H_2

CO_2 —————————— Carbohydrate

Chemoautotrophic microorganisms have been classified into the following types, based on the type of inorganic compound whose oxidation serves as a source of energy. These are sulphur bacteria, iron bacteria, nitrifying bacteria, hydrogen bacteria etc.

Sulphur bacteria

These absorb either hydrogen sulphide or molecular sulphur from the environment and combine them with molecular oxygen. The resulting energy is used for reduction of carbon. Elemental sulphur in the form of granules is deposited. Sometimes, sulphate may be the byproduct which may be excreted, or it might become a part of the mineral content of the cell.

Iron bacteria

These bacteria oxidise iron compounds and convert them into insoluble substances, releasing energy during the process.

Nitrifying bacteria

There are two types of nitrifying bacteria–nitrite bacteria and nitrate bacteria. Nitrite bacteria use ammonia as the source whereas nitrate bacteria use nitrite ions as the source.

Hydrogen bacteria

These utilise molecular hydrogen as the source and combine it with oxygen releasing energy during the process.

Note : More details regarding the oxidation of inorganic metoblites by chemoautotrophic bacteria is given in the chapter on aerobic respiration.)

PHOTOHETEROTROPHIC MICROORGANISMS

These are regarded as the intermediate forms between phototrophs and heterotrophs. The characteristic feature of these nutritional forms is that the external energy source is light (hence, basically these are photosynthesizers). But they require the input of organic raw materials unlike in the case of autotrophic organisms. As such, these organisms are refered to as photoheterotrophs. A small group of photosynthetic bacteria namely **purple**

non sulphur bacteria are typically photoheterotrophs. Like the purple sulphur bacteria these bacteria also possess bacteriochlorophyll as well as the red and yellow cartenoids, masking the green colour.

Purple non sulphur bacteria absorb organic rawmaterials from the environment but these cannot be directly used as food. Instead, these organic materials only serve as a source of hydrogen and hydrogen released from this process is then used for reducing the carbon dioxide. During the oxidation of organic metabolites to release hydrogen, carbon dioxide is also released and these two combine resulting in the form of carbohydrates. The scheme of food production is as follows.

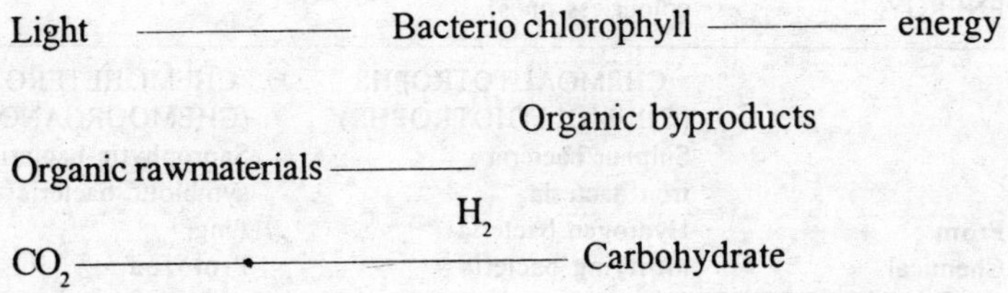

Purple nonsulphur bacteria can manufacture food as shown above only if light is present and oxygen is *absent.*

Another kind of categorisation of the autotrophic microorganisms is possible when we emphasise two criteria (source of rawmaterial – inorganic or organic and source of energy – light or by chemical oxidation) simultaneously. On this basis, we can recognise four categories of microorganisms.

Photolithotrophs
These use light as a source of energy and inorganic materials provide carbon necessary for reduction into carbonhydrates.

Chemolithotrophs
Here, chemicals act as both the source of energy as well as the source of carbon. The energy for reduction comes from the oxidation of inorganic compounds.

Photoorganotrophs
In these microbes, light is the source of energy and reduced organic materials provide the source of carbon.

Chemoorganotrophs
These organisms use organic chemicals as a source of energy as well as directly as food.

The following table gives a summary of the classification of nutritional forms;

	Food Manufactured from inorganic source	absorbed from prefabricated organic source
From PHOTOSYNTHESIZERS Light source PRIMARY ENERGY ENERGY	PHOTOAUTOTROPHS (PHOTOLITHOTROPHS) Purple sulphur bacteria Green sulphur bacteria Microalgae (except colourless ones)	PHOTOHETEROTROPHS (PHOTOORGANOTROPHS) Purple nonsulphur bacteria
From Chemical CHEMOSYNTHESIZERS	CHEMOAUTOTROPHS (CHEMOLITHOTROPHS) Sulphur bacteria iron bacteria Hydrogen bacteria nitrifying bacteria	CHEMOHETEROTROPHS (CHEMOORGANOTROPHS) Saprophytic bacteria symbiotic bacteria fungi Protozoa colourlesss micro-algae
	AUTOTROPHS	HETEROTROPHS

HETEROTROPHIC MICROORGANISMS

These are pigmenetless bacteria and eventhough they absorb H_2O and CO_2 from the environment, they cannot be utilised as a source of energy. Hence, they have to absorb readymade food from the environment and use it directly as food. This means the survival of heterotrophs depends on the presence of Autotrophic organisms.

Majority of the heterotrophic microorganisms are chemoheterotrophs in the sense their source of nutrition is readymade organic substances. Chemoheterotrophs are further classified into holotrophs, saptrotrophs and symbionts.

Holotrophs are free living bulk feeders

They ingest the food and get them digested in their body. Saprotrophs or saprophytes, live on dead and decaying organic remains and absorb the readymade food from them. Many of the bacteria and fungi belong to this category. These microorganisms decompose the organic matter and absorb the nutrient molecules directly. Thus saprotrophs are also called decomposers. As a result of their decomposng activity, saprotrophs constitute a vital link in the global nutriate cycles.

THE PHASES OF GROWTH

The growth of bacterial population may be studied by inoculating a nutrient solution with a small number of microorganisms. When all the nutritional and other parameters are ideal, the cells continue to multiply, until one or the other necessary factors reaches exhaustion and growth gets retarded. If during this period, the culture medium is not replenished with fresh nutrients or waste products (of microbial metabolism) removed, the growth in such a closed system is called **batch culture.** In other words, batch culture may be defined as 'growth of microorganisms in a limited volume of liquid medium, where there is no replenishment of nutrients nor is there any elimination of waste products. Such a closed system (culture) is very helpful in studying the growth characters of microorganisms. When bacteria are transferred to a known volume of culture, their population increase undergoes a characteristic sequence in the rate of increase of cell numbers. This increase obeys the same laws as does the growth of multicellular organisms. *Thus a batch culture behaves like a multicellular organism with its genetically determined stages or phases of growth.*

After the initial inoculation into the medium, increase in cell number may be ascertained by counting them in a specified aliquot at periodic intervals (every hour for the first 24 hours). When the logarithm of the counted cells (viable cells only) is plotted against a growth curve is obtained. Basically, this growth curve is sigmoid (as it is no universally), and exhibits various phases or stages of growth.

Growth curve

A growth curve is the graphic representation of the rate of multiplication of cells plotted against time. For all organisms including microbes the growth curve is sigmoid. There is an initial period which appears to have practically no growth, followed by a spurt in growth. This is followed by a stationery or steady state and finally a decline in the number of viable cells. Based on this behaviour, four stages or phases of growth may be identified. These are– **initial slow phase** or **lag phase; log phase** or **the exponential phase; the stationery phase** and **the death phase** or **the declining phase.**

While the above four phases are clearly recognizable stages, it has to be understood however, that there is no sudden shift from one phase to another in the entire microbial population. The change from one phase to another is gradual, as all bacterial cells will not be in the same stage of growth (except in synchronous cultures). As a result, there will be several intermediate phases between the four standard phases of growth. Taking this factor into consideration growth in microbes may be divided into the following phases :

1. Initial phase (lag phase)
2. Phase of accelerated growth
3. The log phase
4. Phase of decrease in growth rate
5. Stationary phase (the study stage phase)
6. Phase of increase in death rate
7. The declining phase
8. The survival phase

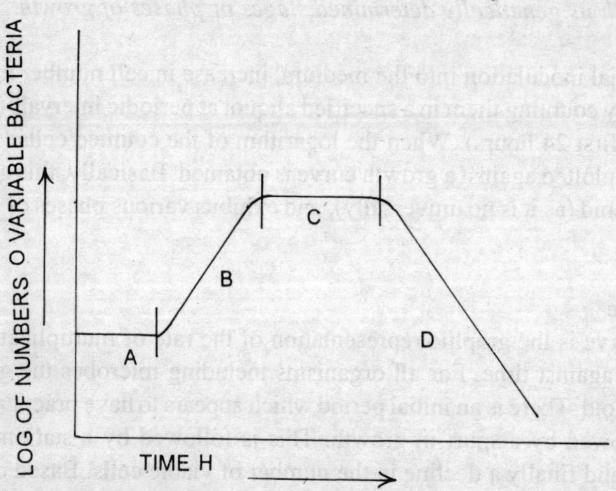

Fig. 2.1 Microbial Growth and Reproduction
The growth curve of microorganisms showing. **A.** Lag phase, **B.** Log phase, **C.** Stationary phase, **D.** Death phase

1. The lag phase

This phase is characterised by the initial period, where there is no increase in the number of cells. The lag phase is said to occupy the time interval between inoculation and establishement of maximum division rate. The actual duration of the lag phase depends on the previous culture history, the age of the culture, and the components of the nutrient medium. During the lag phase, there is an increasc in cell size but there will not be any increase in cell number as there is no multiplication. Nevertheless, it is during this period that the cells are very active physiologically absorbing nutrition from the medium and reaching a phase that prepares them for cell division. Studies have indicated that there is maximum synthetic activity during the lag phase resulting in increase of total proteins, RNA, ribosomes, enzymes etc. There is also an increase in the amount of cellular phosphorous.

The inoculum that has been obtained from the previous culture will have cells that would have adjusted themselves to the growth conditions existing in the old medium. After transfer to the new medium, the cells require a little time to adjust themselves to the new physical and nutritional environment.

It has already been pointed out that the duration of the lag phase depends on the stage of growth of the cells in the inoculum in the previous culture. If the cells of the inoculum are obtained from a culture growing in the log phase, the culture displays very little lag phase in the population increase. On the other hand, if the inoculumis obtained from a culture of dormant cells, the lag phase will continue for a longer period.

On the new culture medium, the nutritional environment many a time requires the synthesis of new enzymes to breakdown specific substrate material. A typical example of substrate induced enzyme synthesis is a phenomenon called **diauxie** (biphasic growth). The appearance of biphasic growth or a double growth cycle is found in media which are complex, and have a mixture of several substrates. For instance in a mixture of glucose and sorbitol, *Echerichia coli* will utilise glucose first, because glucose will induce the synthesise of those enzymes required for its digestion and simultaneously suppress the synthesis of enzymes required for sorbitol utilisation. Sorbitol degrading enzymes are only produced after all the glucose has been metabolised. Hence, in these cultures the two lag phases can be explained by this regulatory mechanisms.

The most notable change in the chemical composition of bacteria during lag phase is the 8–12 fold increase of the RNA concentration.

2. Phase of accelerated growth

This is said to be an intermediate phase between the lag phase and the log phase. At the end of the lag phase after the cells are sufficiently physiologically mature, division sets in. All the cells, however, will not start dividing simultaneously because all of them would not have completed the lag phase at the same time. A large number of cells however would have completed the lag phase. Gradually, however, the cells reach physiological maturity and start dividing On plotting the growth rate, we will observe that the rate of multiplication increase with time. Finally, the rate of multiplication reaches a maximum at the end of this phase. This is said to be the transition phase between the lag and the log phases. It has been observed that during the transition phase the cells are extremely sensitive to alterations in growth conditions like temperature, osmotic pressure and disinfectant chemicals.

3. The log phase (exponential phase)

This phase is characterised by the division of the cells at a constant rate. This constant and steady division results in the doubling of the population. At this stage, the log of the number of cells plotted against time will show a straight line indicating a geometric increase. The cells are said to be in a balanced growth stage, and the mass and volume of the cells increase by the same factor in a manner that the composition of the cells and the concentration of metabolites remain constant for some period of time.

During this phase, cells are actually smaller in size as they are constantly dividing. Physiologically, the cells are said to be very young and actively multiplying. With reference to the physiological activity, the entire population of cells seem to be in a uniform state.

The multiplication of the cells and the doubling time actually depends on the organism in question and the prevailing growth conditions. Enterobacteria can double in a time of 15–30 minutes; *E.coli* growing at a temperature, 37°C can double in 20 minutes. On the otherh and *Nitrosomonas* and *Nitrobacter* have a doubling time of 5–10 hours.

The log phase is a most suitable period for studying biochemical and physiological properties of cells, because in most cells the metabolic activity will be at the peak. Like in the previous phase, the cells are extremely sensitive to the influence of environmental factors such as pH, temperature, aeration, concentration of substrates etc.

4. Phase of decrease in growth rate

During this phase, the rate of cell division comes down and the cells do not multiply with the same rapid sequence as during the log phase. The retardation of the rate of cell division is due to a number of factors. The most important being the gradual decrease in the quantity of available nutrients and the accumulation of toxic waste (product of microbial metabolism). This is a transition phase before the cell division stops completely.

5. Stationary phase (the steady stage phase)

This phase is said to begin when the cells can no longer divide as the amount of available nutrient has almost reached the minimum. The growth rate of the culture gradually gets decreased before all the nutrients get exhausted. In this phase, a high population from the limitation of substrate, other factors such as very high concentration of cells, low partial pressure of oxygen and accumulation of toxic waste products also lead to the initiation of the stationary phase.

At the cellular level during the stationary phase, reserved food materials get consumed, a proportion of ribosomes may be degraded and enzymes may still be synthesised. A viable population count at this stage shows no change.

In fermentation technology involving microbial metabolism (penicilin production), the stationary phase is the real production phase. In biotechnology, this phase is said to be the *idiophase* (production phase). The bacterial mass that is produced at the time of the stationary phase is known as the yield.

6. Phase of increasing death rate

During this phase ,there is a gradual decrease in the number of the viable cells with increase in time. Due to a number of factors, the death of the cells gradually increases and reaches a maximum at the end of this phase. This is a transition phase between the stationary phase and the declining or the death phase.

7. The declining phase

Also known as the logarthemic death phase, during this phase the number of cells decreases exponentially, that is 50% of the surviving cells die in each successive equal time interval. For example, if a population decreases from 10 lakhs to 5 lakhs in the first hour, in the next hour it decreases to 2.5 lakhs, and in the next hour it comes down to 1.25 lakhs, and so on. Thus, here the rate of decline in the number of cells is constant, and it can be calculated using the same formula as for the calculation of logarthemic increase. The calculation indicates the time required to reduce the population by half. This calculation is very useful while considering the germicidal effect of an agent.

The death phase and the cause of the bacterial death in the normal nutrient media have not been thoroughly investigated (Schlegel, 1993). A variety of conditions, however, are known to cause bacterial death. The death rate of bacteria is not uniform just as in the case of cell increase. Some species such as *Neisseria gonorrhoea,* die rapidly as the cells are very susceptible to autolysis. Autolysis is known to be brought about by enzymes called autolysins.

8. The survival phase

This is also known as the decrease in growth rate. During this phase the rate of the death of the cells ultimately reaches an equilibrium, wherein the rate of growth and the rate of death finally balance each other and a very low population of cells is maintained. This stage is called the survival phase. Survival depends on type of organism and other environmental conditions. Some microbes die off within 4–5 days while others might remain viable for months and even years. The formation of endospore at this stage gives an unique opportunity for the cells to provide for very long periods even after the all vegetative cells are eliminated. This is also a stage where, perhaps the mutant forms induced recently find the conditions favourable for rapid growth and get selected.

Growth curve parameters

When the growth of microbes in a batch culture is followed by dry weight determinations, the growth parameters of interest are the yield, the exponentail growth rate and the duration of the lag phase.

Yield may be defined as the difference between the initial and the maximum bacterial mass. It is expressed in terms of grams of dry weight. Of special importance here is the relationship of the yield to substrate consumption. When both these values are expressed in weight units, the ratio is said to be the yield co-efficient.

The exponential growth rate is a measure of the speed of cellular growth in the log phase. It is calculated from the bacterial concentration at initial and final time.

The lag phase like the exponential growth rate is an important parameter for considering the properties of the organism and the suitability of the medium for the bacterial growth.

Factors influencing the bacterial growth (Refer Chapter 1)

Continuous culture of microorganisms

In a batch, which is the ideal medium for the study of growth phases, the cultural conditions undergo continuous change. The bacterial growth curve rises and substrate concentration comes down. As the bacterial growth curve indicates, there are a number of transitional phases between the standard growth phases indicating not all the cells are in the same physiological condition.

There are cells on one extreme and there are dying cells on the other. For many, physiological conditions and the state of the microbes vary. It is ideal, however, for the purpose of metabolic study of an organism, all the cells are in an identical state of growth. In other words, it is desirable to maintain a microbial population in the log phase of the growth in a constant environment. This can be accomplished by a technique called continuous culture technique wherein, there is constant addition of new growth medium and concomitant withdrawl of equal volume of the bacterial culture. By eliminating a quantity of bacterial population and at the same time adding fresh nutrient medium, the growth state of the population is maintained at a particular stage.

A variety of methods have been evolved to maintain a continuous culture of microorganisms for the purpose of study. Two such devices are the *chemostat* and the *turbidostat.*

Chemostat

The chemostat is a mechanism for the continuous cultivation of the microbes. The system consists of a culture vessel and a reservoir. In the culture vessel, the medium is thoroughly mixed to obtain maximum homogeniety. Fresh nutrient medium is supplied to the culture vessel from a reservoir of sterile medium at a constant predetermined rate. The volume of materials in the culture vessel is kept constant by a device which allows for the withdrawal of some amount of culture medium alongwith accumulated waste products and old or dead cells. At the same time, fresh culture medium is added, so that the level is maintained constantly.

The level of growth is maintained by limiting the concentration of a particular nutrient in the medium. The growth limiting nutrient is added to the medium at such a concentration below that required for maximum growth in a batch culture.

The steady rate of bacterial population in a chemostat is maintained by a delicate balance between the growth rate and the dilution rate. Dilution rate may be defined as the ratio of inflowing medium per hour to the volume of culture. In the chemostat, the maximum stability is attained within a range of dilution rates where the density changes get altered very slightly at low dilution. As a result, the change in the bacterial mass is O and the bacterial concentration remains constant. In other words, the exponential increase in cell mass is balanced by a negative exponential process of the same magnitude.

Calculation of growth in a chemostat

As has been explained above, constant growth is maintained in a chemostat through a balance between outflow and inflow. The rate of loss of cells through overflow can be expressed as

$$\frac{dM}{dt} = \frac{F}{V} = DM$$

Where F=flow rate, V=culture volume. Here, flow rate is measured in culture volumes per hour. The expression F/VM is called the dilution rate (D). In a continuous culture,

KM=DM

K=D

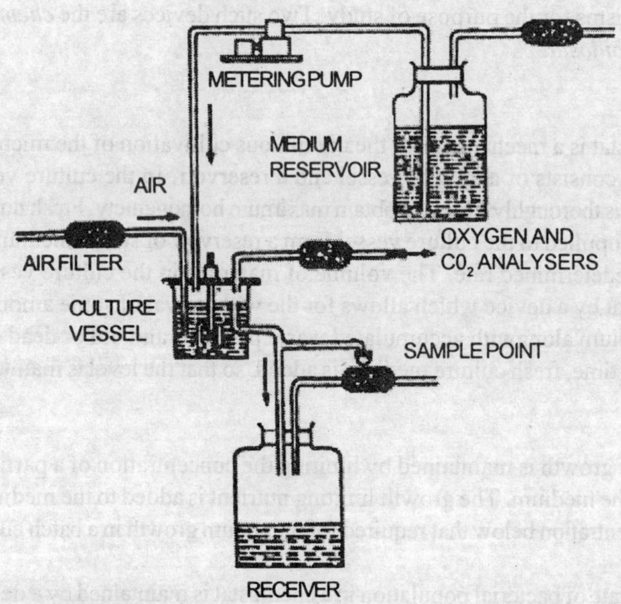

Fig. 2.2 Microbial Growth and Reproduction
Continuous culture of bacteria using a chemostat

In the above, the growth rate (K), equals the rate of dilution (addition of sterile nutrient), in a continuous culture. Under such conditions, the growth of the culture is rather linear than exponential, while the multiplication of individual cells is exponential.

Turbiodostat is another kind of a continuous culture technique. In this, the system has an optical sensor which measures the optical density in the growth vessel. Changes in the density of the culture medium retard the passage of the light through the culture. These changes automatically activate the mechanism that regulates the inflow of the nutrient and the outflow of the waste from the culture vessel. Turbiodostat is a more precise and sophisticated technique than the chemostat as maximum sensitivity and stability are achieved through optical measurements.

Dialysis technique is another device which can maintain the culture in the logarthmic phase for longer periods.

Continuous culture systems prove very beneficial in research studies as they can provide a constant source of cells at a particular stage of growth. Additionally, the system also allows the cells to be grown continuously in limiting concentration of nutrients. Such a kind of growth provides information while studying a particular organism, as to the nature of the limiting substrate. Such a system is very ideal to isolate a specific organism that catabolizes a particular nutrient. Combining this with enrichment cultures, specific microbes may be easily isolated.

Synchronous cultures

These are composed of populations of cells that are in identical stage of growth in their life cycle. In this culture, all the cells in the medium will divide simultaneously; will grow into cells and divide again in the next generation at the same time. Thus, the entire population of microbes is uniform with reference to their growth phases. Synchronous cultures are of great advantage to study the various factors of microbial growth like organisation and differentiation as they provide a large amount of identical material. These analysis are difficult to study in a single microbial cell. Measurements of growth parameters conducted on this population can be simply applied on single individual cells also.

A number of techniques are employed to obtain synchronous cultures. These include physical separation of cells, manipulation of culture, by varying chemical composition etc.

Growth rate and generation time

Bacteria multiply asexually in large numbers mainly by means of binary fission. In binary fission, which is an amitotic division, each cell divides to produce two daughter cells and they in turn divide to produce 4 cells, 8 cells and so on. Thus, the multiplication corresponds to a geometric progression as shown below :

$2^0 - 2^1 - 2^3 - 2^n$ - etc.

In other words, the cells increase as 1— 2 — 4 — 8 —16 etc.

Generation time has been defined as the time required for the cell to divide to produce two cells or for a population to double itself. The generation time or the time required for doubling of the population varies with the type of organism and also with the cultural conditions. The table gives a comparative generation time for some of the microorganisms.

Organism	Temperature (^{0}C)	Generation Time (Min)
Bacillus stearothermophilus	60	11
Escherichia coli	37	20
Bacillus subtilis	37	27
Bacillus mycoides	37	28
Staphylococcus aureus	37	28
Streptococcus lactis	37	30
Pseudomonas putida	30	45
Lactobacillus acidophilus	37	75
Vibrio marinus	15	80
Mycobacterium tuberculosis	37	360
Bradyrhizobium japonicum	25	400
Nostoc japonicum	25	570
Anabaena cylindrica	25	840
Treponmea pallidum	37	1980

Mathematical expression of generation time

When the number of cells in the initial inoculum as well as the growth conditions in the medium are known, it will be possible to calculate the time required for a population to double itself. The following data are necessary to calculate the generation time.

1. The initial number of cells at the beginning of growth.
2. The number of cells present at the end of a specific time period.
3. The time interval.

Using the above data it is possible to calculate the generation time for any given population with the help of a series of equations. If we consider that the experiment starts with a single cell, the total population (*M*) of a bacterium at the end of a given period can be expressed as follows :

$$M = 1 \times 2^n$$

In the above 2^n indicates the quantity of the bacterial populationo after n number of generations. Generally, however, the experiment not given with a single cell, and it has hundreds of cells. Hence, the expression may be modified as follows.

$$M^n = M^0 \times 2^n$$

Here M^0 – number of organisms initially; M^n = number of organisms after n generations and n = number of generations. To solve the equation for n, we should calculate as follows :

$$\log M^n = \log M^0 + n \log_2$$

$$n = \frac{\log M^n - \log M^0}{\log_2,}$$

From the above equation we can calculate the number of generation if we know the initial population and the final population after a specified time T. The generation time G = T (the time elapsed between initial population Mo and Mn.) divided by the number of generations n as follows :

$$G = \frac{T}{n} \quad \frac{T \log_2}{\log M^n - \text{Log } M^0}$$

An organism in a balanced growth would show a rate of increase which is proportional to the number or mass of microorganisms present at that time. The rate of increase of cells is an index of the growth rate andit is called the exponential growth rate constant (K). K is defined as number of doublings of population in unit time and it is usually expressed as the number of doubling per hour. Using this, an alternative method of calculation of bacterial growth may be shown as follows :

$$M^n = M^0 \times 2^{KT}$$

M^n = Population at time T

M^0 = Population at time 0

If we take the logarithms $\log M^n = \log M^0 \times KT \log_2$ and, to solve the equation K we have to calculate as follows :

$$K = \frac{\log M^n = \log M^0}{T \log_2}$$

The above equation indicates the exponentail growth rate constant (K) is reciprocal to the generation time, i.e.,

$$G = 1/K$$

For *E.coli,* the generation time is 30 minutes i.e., there are two doubling for every hour (generation time is usually calculated as number of doublings per hour i.e., if 1/2 = 1/K then K = 2 doulbings per hour).

Methods of measuring microbial growth

Measurement of microbial growth and physiology is an important parameter of microbial study. This measurement to a great extent depends on the quantity (number) of microorganisms present in a culture. Therefore, measurement or quantification of microbial population is a must for any meaningful assessment.

Growth of microorganisms can be quantatively measured using a variety of techniques. These techniques fall into the basics categories mentioned below.

1. ***Determination of cell numbers***

   A. **Total count** (Non viable count)
      - i. Bread smear method
      - ii. Haemocytometer method
      - iii. Proportional count method
      - iv. Electronic counter method

   B. **Viable count** (Indirect count)
      - i. Plate count
      - ii. Membrane filter count
      - iii. Serial dilution count
      - iv. Drop technique
      - v. MPN count

2. ***Determination of cell mass***

   A. **Direct method**
      - i. Dry weight measurement
      - ii. Measurement of cell nitrogen weight

   B. **Indirect method**
      - i. Turbidimetric method

3. ***Determinationo of cell activity***

   A. Measurement of biochemical activity (Indirect method)

TOTAL COUNT (Non viable count)

Total count of microorganisms in any given suspension indicates the total number of cells which includes both living and dead. Hence, it is called a non viable count. The following are various methods of total count.

Bread smear method

In this method, a known volume of (0.01ml) microbial suspension is spread uniformly over a clean, sterile glass slide within a fixed area of a one square cm.

The smear is then fixed to the slide and stained. After placing a cover glass over the smear, it is observed under oil immersion objective and the cells are counted in a microscopic fields. Several microscopic fields are counted and an averages is taken. The total cells per square cm is then calculated by determining the number of microscopic fields per square cm. The total count may then be calculated as follows.

1. Area of microscopic field = πr^2
   r (oil immersion lens) = 0.08mm
   Area of the field under oil immersion lens
   $\pi r^2 = 3.14 \times (0.08\text{mm})^2 = 0.02$ sq cm.
2. Area of smear 1 sqcm = 100 sqmm
   Therefore no of microscopic fields 100/0.02 = 5000
3. Average number of bacteria per field

say 50x5000 = 25,0000 cells per square cm (i.e., number of cells per 0.0001 ml of suspension)

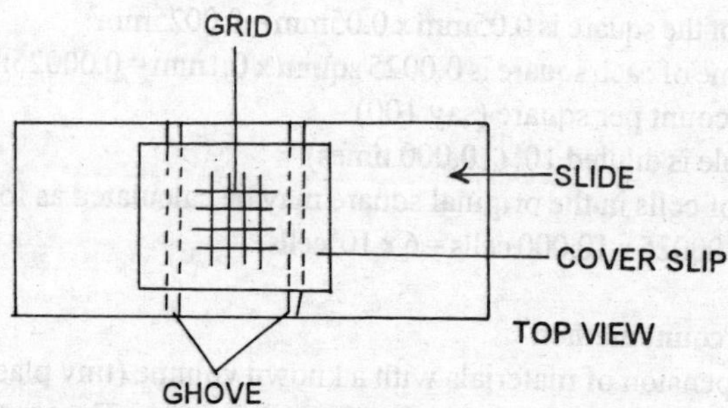

Fig. 2.3 Microbial Growth and Reproduction
Diagrammatic representation of Haemocytometer

Haemocytometer method

Also called counting chamber method or the Petroff - Housser chamber method, this consists of the use of a special microscopic slide. This slide is also called haemocytometer as it was originally meant to count the blood cells.

Petroff-Hausser chamber consists of

1. A thick glass slide which has a depression in the centre having a groove on its both sides.
2. A cove slip

The centre of the depression consists of a specific area divided into 9 squares measuring 1 sqmm. The centre square is further divided into 24 smaller squares each with a side of 0.2 mm. Each of these squares is again divided into 16 squares covering an area of 0.0025 sq mm. Counting is done in this small square under the microscope. The procedure for counting is as follows –

1. Clean both the slide and coverslip to make it free of all dust particles.
2. Place a known quantity of dilute bacterial suspension with the help of a pipette syringe.
3. Place the coverslip firmly on the suspension.
4. Allow the suspension to settle. Place the slide on the stage of the microscope.
5. Count the number of cells in the small squares. Take an average of 50 squares.

Calculation of cell number

(a) The area of the square is 0.05mm x 0.05mm = $0.0025mm^2$

(b) The volume of each square is 0.0025 sqmm x 0.1mm = $0.00025mm^2$

(c) Average count per square (say 100)

(d) The sample is diluted 10^4 (10,000 times)

The number of cells in the original square may be calculated as follows

100 x 1/0.00025 x 10,000 cells = 6×10^6 cells.

Proportional count method

Standard suspension of materials with a known volume (tiny plastic beads) is taken and mixed with an equal amount of cell suspension. The mixture is spread on the slide and stained. The particles and cells in the microscopic field are counted. A number of fields are counted to take the average. If an average of 5 particles and 20 cells per field is obtained, if the number of particles is one ml of standard suspension is 10,000, then the number of cells per ml of suspension is 20/5 x 10,000 = 40,000 cells/ml.

Electronic counter method

In this method, a special instrument called the coulter counter is used for direct counting of cells from a suspension. The instrument can count thousands of cells in a few seconds. But this system has a disadvantage in that, the counter counts even dust particles. Hence, the suspension must be absolutely free of any foreign particles.

Advantages of total (nonviable) **count**

1. Simple and rapid
2. Morphology of cells also may be studied

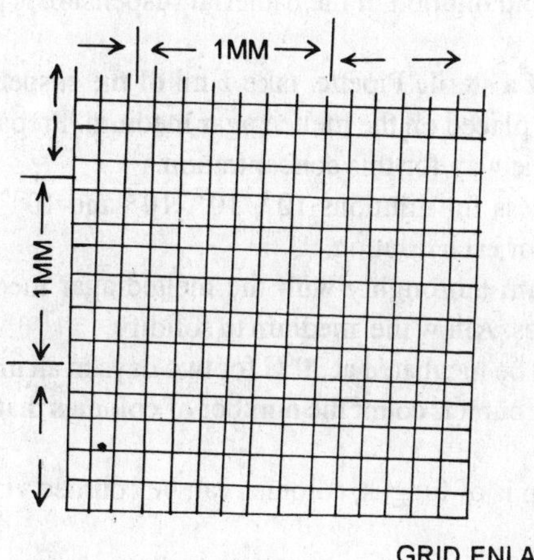

Fig. 2.4 Microbial Growth and Reproduction
A haemocyometer magnified

Disadvantages

1. Includes living and dead cells
2. Accuracy might decline, if the suspension is dense
3. Statistical errors are possible

VIABLE COUNTS

In viable counts, only the living bacterial cells are taken into consideration. The principle behind viable count is that all viable cells or spores, under suitable growth conditions multiply and that (as in an agar medium) each cell or spore forms a colony. The number of colonies therefore is the same as the number of viable cells present in the original sample.

In viable count methods, a knwon quantity of diluted sample is taken in such a way that the number of colonies that develop on an agar plate will be in the range of 30–300.

Viable count techniques are of the following types.

Standard plate count

Also called the spread plate technique, in this method appropriately diluted suspension is inoculated on a series of agar petridishes. The procedure is as follows.

1. A series of ten fold dilution of the bacterial suspension is prepared (10^2 to 10^{-10}).
2. With the help of a sterile Pipette, take 1 ml of the suspension and allow 0.1ml of it to be placed on the melted agar medium. Prepare some replica plates in the same way for this concentration.
3. Repeat the process for dilutions 10^{-4}, 10^{-6}, 10^{-8} and 10^{-10} also, use fresh sterile pipettee for each dilution.
4. Mix the inoculum thoroughly with the melted agar medium by slowly rotating the plates. Allow the medium to solidify.
5. The plates are to be incubated at 30°C for two days in an inverted position.
6. Using a colony counter, count the number of colonies that develop on the plate.
7. If the suspension is of fungus, colonies can be counted visually without a counter.

To facilitate easy counting, the plate may be divided into four sectors by marking on the underside of the plate using a glass marking pencil.

Calculation

For the purpose of calculation, a specific dilution level (say 10^{-4}) may be taken as colony number varies with each dilution. A plate with colonies ranging between 100–300 is deal for the purpose of calculation.

If dilution 10^{-4}, has 200 colonies developed on the plate, the number of cells in the original sample may be calculated as follows.

$$200 \times 10,000\ (10^{-4}) = 20,00,000 \text{ cells } 1/1\text{ml} = 2 \times 10^6 \text{ cells/ml}$$

As each colony develops from a single cell in the original suspension, the number of cells must be 2×10^6.

The plate count has some disadvantages. If the original suspension has a mixture of microbes, not all of them may grow properly on a standard medium. If the suspension contains aggregates of cells, the number of colonies developed may not exactly indicate the number of cells, as many cells may develop into a single colony.

The plate count method is used for the estimation of microbial population in milk, water, soil etc. Plate count method is, hov.ever, eminently suitable to count the living cells.

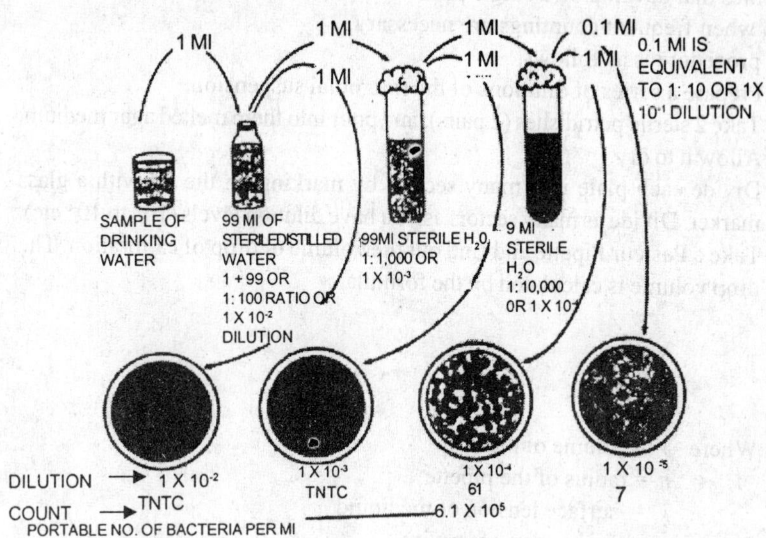

Fig. 2.5 Microbial Growth and Reproduction
The viable count technique (standard plate count)

Membrane filter count
This method follows the same principle as in a standard plate count. A diluted suspension of microorganisms is filtered through a millipore filter. The microbes are retained on the filter disc. The filter disc is placed on a culture medium in a petridish. The plates are incubated and the colonies are counted on the filter disc. Following the same method of calculation given for standard plate count, the number of clells may be determined.

Membrane filter method has many advantages over the plate count. A large volume of sample may be analysed when organisms are few in number. This method is better when a variety of microbes are to be analysed. Filter discs with microbes may be kept in different nutrient media suitable for different microbes.

Serial dilution method
In this method, a suspension of microbes is diluted sequentially and plated on agar medium. (For details see Chapter 1. Culture of microbes).

Drop technique

This is an ideal method where a comparison can be made of the number of colonies that develop on a single plate with different dilutions. This is also of help when frequent countings are necessary.

The procedure is as follows.

1. Prepare a series of dilutions of the microbial suspension.
2. Take 2 sterile petridishes (2 pairs), and pour into them melted agar medium. Allow it to dry.
3. Divide each plate into many sectors by marking on the lid with a glass marker. Divide as many sectors as you have dilution levels (10^{-1} to 10^{-6} etc).
4. Take a Pasteur Pipette and find out the volume of drop of calibration. The drop volume is calculated by the formula.

$$V = \frac{2rj}{dg\,\theta}$$

Where V = volume of the drop
r = radius of the pipette
j = surface tension of the liquid
g = acceleration of gravity
θ = correction factor of the neck of the drop

In the above, r and g are constant, and v and θ are almost constant. The value is affected by the volume of j. To keep the value of j constant, the inner side of the pipette must be kept absolutely clean. In order to calibrate, such a quantity of liquid, deliver the solution drop by drop into a measuring cylinder. Count the number of drops required to make the volume to one ml. Suppose, 100 drops make one ml then, each drop has 0.01ml

5. Arrange the plates, take a particular dilution, the lowest and put a drop of the suspension on the sector marked 10^{-1} (by lifting the lid) dilution. Thoroughly clean the pipette (or use a separate pipette) and repeat the process for other dilutions. (10^{-2}, 10^{-4}, 10^{-5} and 10^{-6})
6. Allow the drops to settle and dry. Invert the plates and incubate them at 30°C for two days.
7. Count the number of colonies, as follows.

The volume of each drop is 0.01ml., if the sector at dilution 10^{-4} has 50 colonies, the original sample should have ($50 \times 1/0.01 \times 10^{4}$) 5×10^{7} cells/ml.

Most probable number (MPN) count

Also known as the multiple tube method, it is used for estimating the number of viable bacterial cells in a liquid medium. In this method, a set of culture tubes are inoculated with a fixed volume of the suspension (say 5ml for first tube, 4ml for second, 3ml for third tube etc), with the series of tubes getting progressively smaller volume of suspension. Tubes are incubated and observed for bacterial growth. It is assumed that growth will take place in any tube which receives at least one living cell. It should be understood that if the original sample has very few bacteria, then very few series of tubes show positive growth as the series with greater dilution will dilute them out.

For instance, three sets of culture tubes may be taken as follows :

First set - 1ml inoculum

Second set - 1ml inoculum

Third set - 0.1 ml inoculum

The above is said to be triple dilution series. The quantity of the culture medium, however, is same in all the tubes. After incubation, if

1. All the three tubes of first set show growth
2. Only two tubes of second set show growth
3. Only one tube of first set shows growth

The series may be expressed as 3–2–1. This can be checked for MPN from a standard table of MPN. The number may also be calculated as per the Thomas formula.

$$\text{MPN/100ml} = \frac{\text{Number of positive tubes}}{\sqrt{(\text{ml of sample in negative tubes}) \times (\text{ml sample in all tubes})}}$$

DETERMINATION OF CELL MASS

This is a method in which, the weight or mass of the cells is estimated as an indicator of increased growth. There are two methods of measuring the mass direct method in which the cell mass is actually weighed, and an indirect method in which cell mass is determined indirectly as a function of optical density (turbidimetric method).

Direct method

1. Measurement of dry weight

This is a simple and direct method of measuring the cell mass. The culture suspension is centrifuged and the pellet is repeatedly washed to remove all foreign particles. After drying, the pellet which consists of cells is weighed. This method is useful in large scale operations such as industries, where a bulk amount of microbial growth is involved.

2. Measurement of cell nitrogen

A major chemical constituent of cell is protein, of which nitrogen is an important ingredient. A microbial growth can be assessed in terms of the quantity of nitrogen content. The cells are to be obtained by centrifugation as mentioned above, and the cell nitrogen is estimated by chemical analysis. This method is useful with dense cell suspensions where the amount of growth is large.

Indirect method

Turbidimetric method

This is an indirect method of measuring cell mass by measuring the varying intensity of light as it pásses through a cell suspension. The apparatus used for this is a photocolorimeter or spectrophotometer. A colonimeter functions on the principle, that subject to certain limitations, the light absorbed or reflected by a bacterial suspension is directly proportional to the concentration of cells in culture. Thus, by applying nepholometry (measurement of reflected rays), or turbidimetry (measurement of percentage light absorption). (For more details see Volume I College Microbiology by the same author).

For turbidimetric estimations, a photocolorimeter is used. This instrument has a source of monochromatic light (light of a single wavelength), which is facilitated by a filter that allows only the desired wavelength of light to be transmitted. The light ray passes through the culture and the amount of light transmitted is measured by means of a photoelectric cell connected to a glavanometer. The percentage of light transmitted/absorbed is usually expressed in terms of optical density (OD), which is directly proportional to the cell concentratin. OD is a function of the negative log of the percentage transmission and is expressed as 2-logG. Therefore OD = log 100 - log of galvanometer reading.

The value of OD reading of a bacterial suspension is only a relative value in that it does not directly indicate the number of cells/mass of bacteria. In order to interpret the OD value in terms of cell mass/number, it has to be compared with a standard curve which indicates the direct relationship of OD and mass.

A standard curve can be prepared by obtaining a series of dilutions of a bacterial suspension. A suitable aliquot from each suspension is then taken and the cell mass (dry weight) or number is measured by plate count or any suitable method. The same dilution level is now used to measure the OD. By measuring the OD and corresponding cell mass/number for series of dilutions, a direct relationship between the two sets of data may be established. This is done by plotting the data on a graph (mass/number on X axis and OD on Y axis), wherein a linear standard curve is obtained.

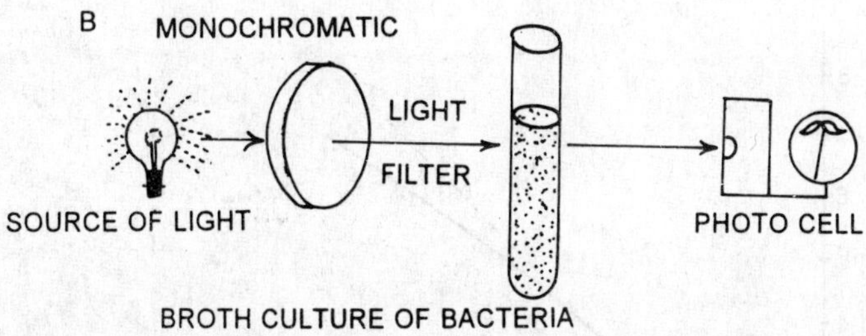

Fig. 2.6 Microbial Growth and Reproduction
Method of turbidimetric estimation of bacterial population

Cell mass/ number of an unknown sample can be determined by taking the OD and compmaring it with the corresponding value on the standard curve. Turbimetric measurements, however, are very accurate only with moderate density. If a suspension is very dense it has to be diluted and the dilution factor must be taken into consideration while calculating cell mass.

Disadvantages

1. If the suspension contains any colouring matter, it effects light transmission and the absorbancy of light is not just due to cell concentration.
2. Turbidity measurements are not useful for viable counts as they measure both dead and living cells.

DETERMINATION OF CELL ACTIVITY

Metabolic products produced from a bacterial cell mass can be quantified and expressed in terms of cell mass quantity. The logic behind this is, more number of cells the more the quantity of metabolic product in question. This is an indirect method of assessment of cell mass. For instance, the amount of acid produced in a medium of cell suspension is a measure of the concentration of cells.

Sometimes concentration of specific enzymes also may be assayed to find out the cell mass.

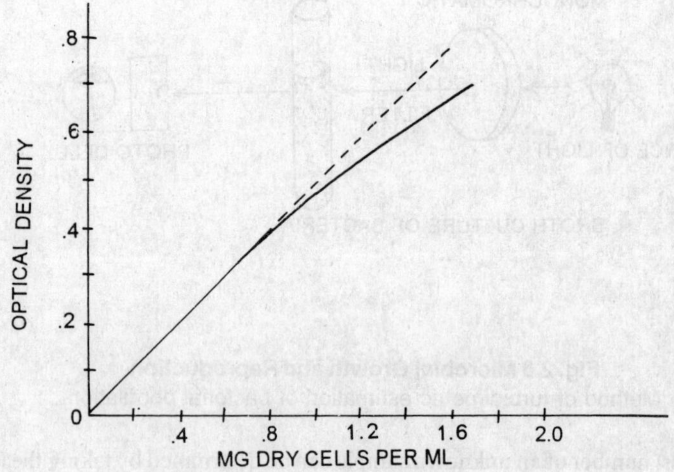

Fig. 2.7 Microbial Growth and Reproduction
Graph showing standard curve and dry weight curve using photocolonimetric estimation (broken line represents standard durve)

REPRODUCTION IN BACTERIA

Bacteria reproduce asexually as well as sexually. Asexuall reproduction takes place by the following methods. 1. Fission (cell division) 2. Budding 3. Endospore formation.

Cell division Generally bacterial cells do not divide mitotically as is true for all prokaryotic cells. There are three aspects of cell division called binary fission. These are 1. DNA duplication 2. DNA partitioning 3. Cross wall formation.

The cell division is completed by the doubling of all the cell components and their precise distribution and partitioning between the two daughter cells. Unlike in eukaryotic cells, there is no spindle formation and no resolution of chromatin into chromosomes.

Electron microscopic studies have revealed the following details in bacterial cell division.

DNA duplication. The two strands separate and each strand then replicates a new strand.

DNA partitioning. This involves the distribution of DNA between the two daughter cells. As the DNA is free and not condensed into chromosomes, it is a complicated task to assure equal distribution of genetic material between the two daughter cells. It is belicved that (Kleppe et al, 1979) the DNA molecule attaches itself at some point to the plasma membrane and after duplication, one strand swings towwards one half of the cell keeping itself in the same attached position. Both the havles of the cell receive one double stranded DNA each. The attachment of DNA to the plasma membrane is to assure proper distribution without entanglement.

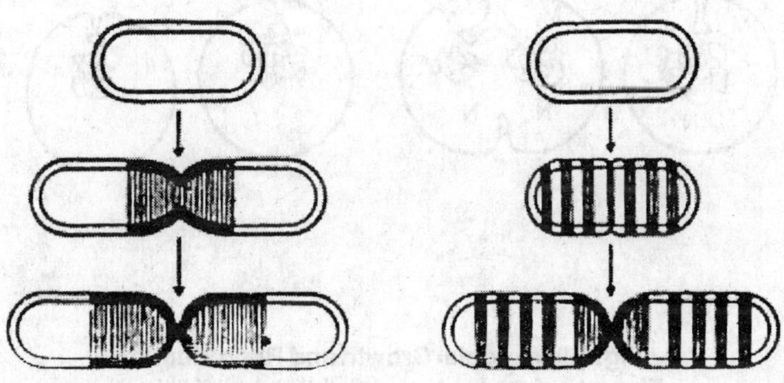

Fig. 2.8 Microbial Growth and Reproduction
Diagrammatic representation of wall formation in gram positive (left) and gram negative (right) bacterium

Some recent studies have indicated that mesosome present in the central portion of the cell plays a significant role in the distribution of DNA to daughter cells. Electron micrographs have shown the attachment of DNA to the mesosome. Along with DNA, the mesosome also bifurcates and each bit of mesosome carries along with it a double strand DNA molecule that ultimately goes to a daughter cell.

Cross wall formation

The cytoplasmic membrane of the bacterial cell grows inward in the middle part of the cell. Mesosome (Chondriod), an intracellular membranous structure continuous with the cytoplasmic membrane plays an importnat role in the formation of septum. The mesosmes of gram positive bacteria are more distinct and may appear as parallel or concentric lamellae, or as vesicles or tubules which get interconnected. Studies have indicated to the contribution of mesosomes towards the synthesis of new membrane. This is followed by the growth of the cell wall and finally the parent cell gets divided into two daughter cells.

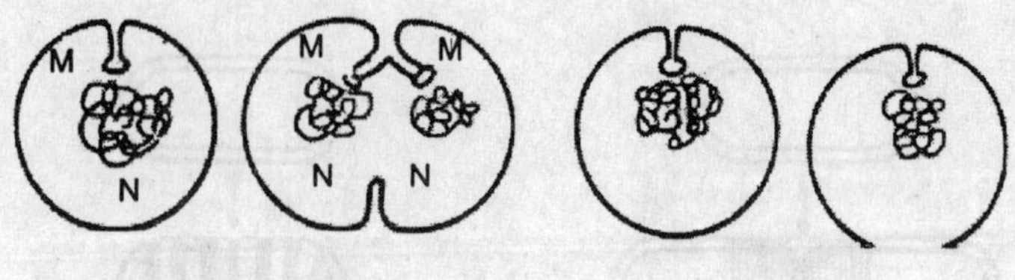

Fig. 2.9 Microbial Growth and Reproduction
The role of mesosome in cell division of bacteria

The formation of the septum starts below the equatorial ridge in the cell wall. The newly synthesized cell wall is plastic enough to undergo changes in shape to achieve the final configuration. Hydrolase enzyme is known to bring about the separation of two daughter cells in *Bacillus subtilis* and *Staphylococcus facalis* (Peberdy, 1980).

Mesosomes of gram negative bacteria are smaller and less intricate. They are often seen as small invaginations of the cytoplasmic membrane. Cross wall formation has been studied in detail in *Escherichia coli,* a typical gram positive bacterium. Here, the cell consists of a middle peptidoglycan layer between outer membrane and cytoplasmic membrane. At the site of formation of septum a bud or fold originates very close to the mesosome. This is followed by the invagination of peptidoglycan and cytoplasmic membrane. Continuation of the invagination process ultimately leads to the formation of the cross septum.

Budding

In the process, a protrusion first develops, into which cytoplasm migrates. The nucleus splits into two and one bit cnters the protrusion. This is later cut off from the cell by a cross wall. This is the bud. At maturity, this separates from the parental cell and develops into a new individual.

Endospore

Many of the bacterial like *Bacillus, Clostridium* etc., form endospores at certain stages in their life cycle. Strictly, endospore formation cannot be called reproduction because only one spore is formed per cell. Endospores are highly resistant to environmental conditions and may be regarded as an attempt by bacteria to withstand the adverse environmental conditions. Endospores are so highly resistant to desiccation, chemicals, temperature and radiation that they are vialable even for centuries. Some ensospore even withstand boiling temperature also for a long time.

As the spores are formed inside the parental bacterial cell, they (spores) are called endospores. On germination, each endospore develops into a single bacteria cell. The spore wall assures the endospore high resistance due to the presence of compounds like diplolinic acid, calcium and a very low water content.

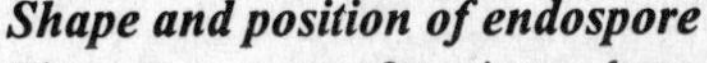

Shape and position of endospore

The spores are of various shapes and occupy various positions in the cell. They may be spherical or oval in shape, and in position may be terminal, sub-terminal or central. They may be bulging giving a swollen appearance to the cell, or non-bulging conforming to the outer parameters of the cell.

Ultra structure of endospore

The outermost layer of the endospore is the exosporium which is thick and wrinkled. Chemically, it is made up of lipids and proteins with a low content of methionine and cysteine. Internal to the exosporium are two spore coats–the outer and the inner. The spore coats may be smooth or show various types of ornamentations like ridges, grooves etc.

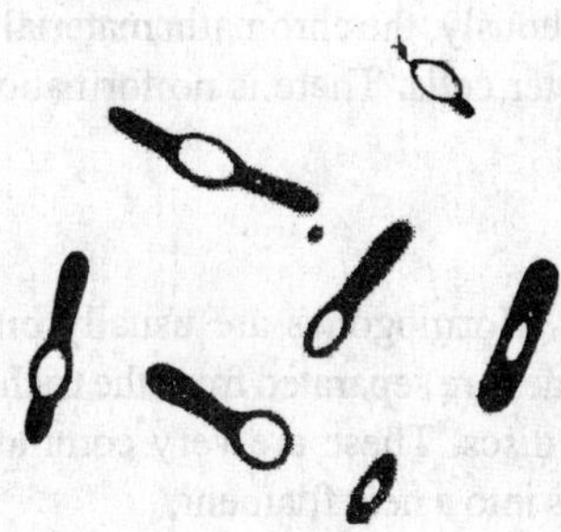

Fig. 2.10 Microbial Growth and Reproduction
Various positions of endospores within the bacterial cells

Chemically, the spore coats are made up of lipids, proteins and glycopeptides.

The protein of inner spore coat has a high quantity of sulphur containing amino acids. The spore coats surround a space called cortex which can be divided into outer cortex and inner cortex. Chemically, cortex consists of dipicolinic acid, peptidoglycan and calcium ions. The cortex is internally lined by the spore wall internal to which lies the cell membrane. Chemically, the membrane consists of peptidoglycans but no teichoic acid. The membrane surrounds the central core of the endospore. The core consists of nuclear material and cytoplasm. This region has DNA, RNA and proteins.
Sexual reproduction as shown by genetic recombination is explained in detail in Chapter 6.

REPRODUCTION IN CYANOBACTERIA

While generally one cay say that there is no sexual reproduction and no gametes are formed, there are, however, specific instances of genetic recombination indicating a type of sexual reproduction much the same way as in other bacteria. Genetic recombination has been demonstrated in *Anacystis nidulans* (Kumar. H.D. 1962), and *Cylindrospermum majus,* and other filamentous forms (R.N. Singl & R.N. Sinha 1965).

The most prevalent methods of reproduction are vegetative propagation and formation of non motile spores.

Vegetative propogation

1. Cell division

The cell division starts with the formation of a ring like outgrowth which grows inwards dividing the cell into two. Simultaneously, the chromatin material also divides and is distributed to the two daughter cells. There is no formation of chromosomes or spindles.

2. Hormogones

This is formed in the filamentous species. Hormogones are usually bits of trichomes consisting of few cells. Hormogones are separated from the trichome by the formation of biconcave separation discs. These are very common in Oscillatoriaceae. Each hormogone develops into a new filament.

3. Hormospores or Hormocysts

At the time of formation of Hormocys't a large amount of food is accumulated in the harmogones and they become thick walled. These are usually formed at the tips of the trichomes. They germinate to form a new filament.

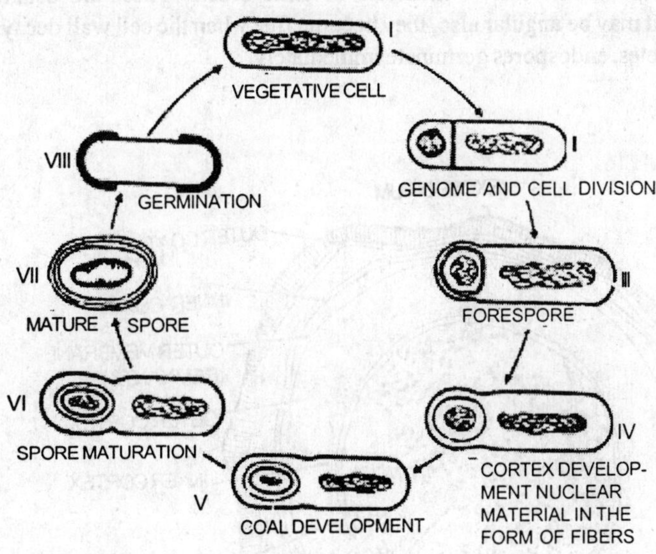

Fig. 2.11 Microbial Growth and Reproduction
The sporulation cycle in an endospore forming bacterium

Spore formation

1. Akinetes, Resting spores or Arthrospores

These are non motile spores generally formed in nostocaceae and Rivulariaceae to tide over the adverse environmental conditions. At the time of formation of an akinete' the cell accumulates large amount of food and the cell wall becomes very thick. The wall may be distinguished into outer exospore and inner endospore. This type of spore in which the original cell wall forms the outer most layer is known as "Akinete". Akinetes may be isolated in a trichome or may be formed in a series. These are extremely resistant to environmental hazards and remain viable for long periods. On germination, an akinete develops into a new individual.

2. Endospores

These are also called gonidia and are produced in all the genera of chamaesiphonales and in few members of other orders. These are usually spherical but may be angular also, they become free when the cell wall decays. Unlike akinetes, endospores germinate immediately.

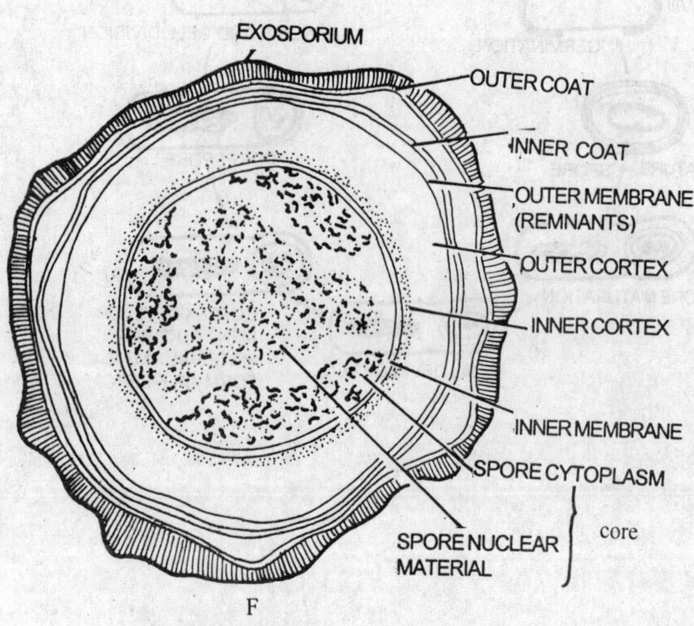

Fig. 2.12 Microbial Growth and Reproduction
Ultrastructure of endospore in *Bacillus cereus*

3. Exospores

In the epiphytic member *Chameosiphon,* exospores are generally formed. After a ceertain period of growth the protoplast abstricts externally cutting off a spore much the same way as a conidium is formed in fungi.

4. Nannospores

In certain forms like *Gloeocapsa,* where the cell division is very rapid the daughter cells are small and are called Nanocysts or Nannospores. They are some what similar to endospore. Eventually, Nanospores develop into new individuals.

5. Heterocysts

All filamentous Cyanobacteria except Oscillatoriaceae have special thick walled cells called Heterocysts. They are formed from the vegetative cells. They may occur singly or in chains. In some general like *Rivularia,* they are always basal while in others like *Nostoc,* they are inercalary.

Each heterocyst consists of a thick two layered wall. At the poles, the wall swells to form a nodule known as the polar nodule. Depending on the position, there may be one or two nodules; one in basal and two in intercalary. As the heterocyst matures, the protoplast inside becomes more and more transparent due to its transformation into a homogenous viscous substance.
Various views have been put forward to account for the nature of heterocysts. One view is that the heterocyst is spore like. Brand (1933), Geitler (1911), Canabouse (1929), and Desikachary (1946), have reported exceptional cases where heterocysts have germinated to give rise to new individuals (Nostoc, Anabena etc). Spratt (1911), has shown heterocysts forming a number of endospores as in *Anabena cycadicae.* These show that heterocysts are archaic reproductive structures which are largely functionless now.

Although functionless, heterocysts have taken a number of secondary functions. They are

1. Helping in the development of Akinetes (Fristch, 1951).
2. Help in the breaking up of a trichome (Kohl, 1903).
3. In genera with false or true branching a relation exists between the origin of the branches and the position of the heterocyst.
4. Help in N_2 fixation.

Genetic recombination

It is generally believed that Sexual reproduction is absent in cyanobacteria, but genetic recombination has been reported in several members. According to Kumar (1962), biochemical recombination seen in the cyanobacteria is a kind of parasexual phenomenon and it differs from true sexuality in that "it is not attended by syngamy or meiosis", but the true function of sexuality via., gene recombination is achieved. Gene recombination seen in cyanobacteria is similar to what is seen in other bacteria.

Genetic recombination was first demonstrated in *Anacystis nidulans,* (Kumar 1962) *Cylindrospermum majus* (Singh and Sinha, 1965), *Anabena dolislum* etc. Genetic recombination has been indicated with reference to antibiotic resistance and such other markers. Kumar (1962), cultured strains resistant to streptomycin and penicillin on separate media, and obtained a resistant variety as shown below :

Strain 1 - S+P-
Strain 2 - S -P+

Strain 1 is sensitive to penicillin, but resistant to streptomycin. Strain 2 is sensitive to streptomycin but resistant to penicillin. When these strains were kept on a medium containing both streptomycin and penicillin, surpirsingly some cells survived and they possessed resistance to both the antibiotics (S+P+). This is possible only by the recombination of strain 1 and strain 2. Pikalek (1962), raised an objection to the experiment mentioned above by staining that penicillin used by Kumar (1962), was not a stable one.

Brazin (1968), was able to demonstrate unequivocally gene reco.nbination in *Anacystis, nidulans* using single resistant strains with reference to polymixin B and streptomycin. He could isolate double resistant strains to both polymixin B and streptomycin.

Nothing definite is known about the mechanism of genetic system involved in gene recombination. According to Kumar (1982), it is likely that recombination is brought about by conjugation between donor and recipient cells much the same way as in other bacteria.

Lazaroff and other Vishniac (1962), have claimed that under certain conditions, filaments of *Nostoc muscorum* actually fuse, much in the same way as heterokaryotics do in fungi.

3

ENZYMES

A plant cell is virtually a microchemical factory in which hundreds of chemical reactions are taking place at amazingly high speeds and that at normal temperature ranges. These biochemical reactions provide sustenance for a plant to absorb, assimilate, synthesize, degrade, grow and reproduce which are essential for survival. Many of these reactions are simple, but most of them are complex, and involve several sequential steps in order to produce a particular product. Conversion of one compound to another has to be precise and exact, there being no excuse for mistakes. For a long time biologists and biochemists wondered, as to how biological reactions occur *in vivo* in such precise manner and at such rapid rates without involving undue increase of temperature. Very soon they got an answer, when Buchner (1987), accidentally discovered that a juice extracted from a yeast cell can perform fermentation as the living yeast cells can do. This means then that the juice contained some substance which can bring about fermentation. The chemical substance bringing about fermentation was soon isolated and given the name zymase. This started the glorious story of enzymes. Enzymes can be described as *Chemical middlemen* mediating and accelerating a variety of biological reactions. It may not be an exaggeration to state, virtually no metabolic activity in living organisms is participating without enzymes. Each reaction requires the participation one two or sometimes many specific enzymes. What are these enzymes? What is their role in metabolism? How do they accelerate chemical reactions? We shall try to find answers to these in this chapter.

The word enzyme is derived from Greek, meaning – *En* = in; zyme = leaven or living. The term was first coined by Kuhne (1978), while working on fermentation, even though it was Buchner who first made an extract of enzyme from yeast cells. Robert Sumer (1926), was the first scientist to purify and crystallize an enzyme revealing the proteinaceous nature. Even though the scientific discovery of enzymes may be comparatively recent, their use for practical purposes was known to early Greeks and Indians. The *Arthashastra* of Kautilya (4th century BC), mentions the use of extracts of barks in certain trees helpful in the process of fermentation.

Definition of enzymes

(a) Biocatalyst;

(b) Organic catalysts;

(c) Specific proteins, simple or compound in structure acting as specific catalyst;

(d) Biological middlemen.

Structure and chemical composition

All the known enzymes are proteinaceous. These proteins called functional proteins, have a very complex structure and have molecular weights sometimes ranging up to several millions. Even some of the simple enzymes like *pipsin* (35,000), *urease* (4,83,000) etc., also have high molecular weights. While all enzymes are made up of proteins, they also have a non-protein part. The bulk of the enzyme, however, is proteinaceous.

The total eyzyme is given the name *holoenzyme* (Protein + non protein), while the protein part is called the *apoenzyme* and the non-protein part is given the name *prosthetic* group. When the prosthetic group is not tightly bound to the apoenzyme, it is called a *coenzyme* or *cofactor.* The presence of a non-protein part in the enzyme is a must as otherwise the enzyme will be inactive. The following are some of the structural attributes of the enzymes.

(i) Apoenzyme

The protein part constitute the major part of the enzyme molecule. The complex proteins that make up the apoenzyme have specific sequences of aminoacids contributing to the specific reactivity of the enzymes. As the proteins are colloids, they offer a large area in relation to their volume. Very few enzymes are wholly made up of proteins. E.g. *proteolytic enzymes.*

Active site

All enzymes possess an area in their molecular organization where substrate materials bind themselves in order to undergo chemical change. This binding site is called the active site of an enzyme. An enzyme may have one or more active sites. In addition to the active sites an enzyme may also have *regulatory sites,* to which regulatory substances bind and regulate the activity of the enzyme.

Prosthetic group

Most of the enzymes possess a non-protein part tightly bound to the protein part known as the prosthetic group. Many metallic ions like Cu, Zn, Mo etc., constitute the prosthetic groups. Organic compounds such as cytochromes, flavoproteins, pyridoxal phosphate, biotin etc., are also known to be prosthetic groups of certain enzymes.

Coenzymes and Cofactors

Strictly speaking, there is no distinction between a prosthetic group and a coenzyme except that the latter is loosely bound to the enzyme molecule. The term cofactor is employed for inorganic ions such as Cl^- Mg^{++} etc., while organic molecules are called conezymes.

Distribution of enzymes in plant cells

Enzymes are found distributed all over the plant cell. But there is a qualitative distribution i.e., not all enzymes are found in all the areas. Enzymes have a localized distribution with reference to the functions they have to perform. The distribution of some important enzymes are listed below.

1. **Nucleus** : *Polymcrases, Transferases* etc.
2. **Chloroplast** : *Carboxydismutases, Phosphatases, Kinases, Dehydrogenases* etc.
3. **Mitochondris** : Enzymes of respiratory pathway - *Dehydrogenases, Cytochrome oxidases etc.*
4. **Cytoplasm** : *Aldolsases, Isomerases, ATP ases, Phosphorylases, Transphosphorylases* etc.

(The list is not comprehensive)

Nomenclature and Classification

The name of the enzyme is usually descriptive and consists of two parts – the first part denotes the substrate on which the enzyme acts, while the second part indicates the type of action of enzyme. For example, *Isocitric dehydrogenase* indicates the enzyme acts on Isocitric acid and removes hydrogen from it. In some of the older nomenclature, this is not followed and the enzymes are named arbitarily. For Eg., *Pepsin, Trypsin, Chymotrypsin* etc.

According to the commission on Enzymology of the International Union of Biochemistry, all enzyme names should end with ase.

Enzymes are classified into six major divisions. These are –

1. Oxidoreductases

2. Transferases
3. Lyases
4. Isomerases
5. Ligases (= synthetases) and
6. Hydrolases.

1. Oxidoreductases

Enzymes which bring about oxidation reduction reactions are included under this group. This group is further subdivided into the following.

(i) *Oxidases* : Reactions involving oxidation of the substrate with molecular oxygen serving as the electron acceptor belng to this category.

(ii) *Peroxidases* : These catalyze the removal of hydrogen from the substrate which combines with hydrogen peroxide (H_2O_2)

$$SH_2 + H_2O \rightarrow \quad 2H_2O + S$$

(iii) *Dehydrogenases*: Enzymes which bring about oxidation of the substrate by removing hydrogen from it belong to this category. The hydrogen acceptor, however, is not molecular oxygen.

$$SH_2 R \rightarrow \quad S + RH_2$$

(iv) *Reductases* : These catalyze the incorporation of both the atoms of oxygen into the substrate.

(v) *Oxygenases* : These catalyze the incorporation of both the atoms of oxygen into the substrate.

(vi) *Hydrolyases* : These enzymes catalyze the reaction in which only one atom of oxygen is added to the substrate.

(vii) *Catalases* : These liberate molecular oxygen from hydrogen peroxide.

2. Transferases

These enzymes transfer a particular group from one substrate to another. These are further divided as follows : based on the type of group transfer which they catalyze.

(i) *Transketolases* : Transfer a ketone group
(ii) *Transaminase* : Transfer an amino group
(iii) *Transaldolases* : Transfer an aldose group
(iv) *Transphosphorylases* : Transfer an every rich phosphate group
(v) *Transcarboxylases* : Transfer a carboxyl group
(vi) *Hexokinases* : Transfer an ordinary phosphate group
(vii) *Gransglycorsylases* : Transfer of a glycosyl group

3. Hydrolases

These enzymes catalyze the hydrolysis of complex substrates into simpler ones (starch into glucose). These include *Carbohydrases* (hydrolysis of carbohydrates), *Lipases* (hydrolysis of fats), *Phosphatases* (removal of phosphate groups) etc.

4. Lyases

These catalyze the removal of groups from the substrate without the addition of water (non hydrolytic).

Fructose, 1,6 diphosphate → Dihydroxy acetone phosphate +phosophoglyceraldehyde

Other examples of lyases are - *Aconitase, Isocitrase, Fumarase* etc.

5. Isomerases

They bring about isomerization of one substrate to another (intramolecular shifting of atoms)

$$\text{Glucose 6 phosphate} \xrightarrow{\text{Isomerase}} \text{Fructose 6 phosphate}$$

6. Ligases

Also called synthetses, they help in the synthesis of a new compound generally with the cleavage of ATP or other nucleoside phosphates.

Isoenzymes

Enzymes which perform similar functions but having different molecular structures are called *Isoenzymes* or *Isozymes*. In other words, *different molecular species of an enzyme can be called Isozymes.* Isozymes can be identified based on the appearance of different bands on a polyacrylamid 3 gel. For example, the enzyme *lactic dehydrogenase* has five isozymase.

Zymogens

These are enzyme precursors or inactivated forms of enzymes secreted by cells. These can be activated to the normal enzyme state by chemical modification.

Constitutive and adaptive enzymes

Not all enzymes are present in the site of action always. Some are synthesized at the time of reaction and soon get degraded after the reaction. Such enzymes are known as *adaptive enzymes.* The other category – *constitutive enzymes* are present always and participate in reaction whenever required.

Properties of enzymes (Physical and Chemical)

Enzymes are characterized by the following properties.

1. Catalytic properties

All enzymes are catalysts, and as such they accelerate the pace of reaction. Being true catalysts, they do not undergo any chemical change and do not figure in the end products of a reaction. They also do not disturb the equilibrium of a reaction. Being catalysts, they are required in very small quantities. The number of molecules of substrate converted into product per minute is called, the *turnover number* of the enzyme indicating its efficiency. Turnover number of different enzymes varies from 100-3,00,000.

The velocity of enzyme action like that of a true catalyst is proportional to the concentration of the substrate/concentration of the enzyme.

2. Reversibility of action

Like all catalysts, enzymes can accelerate the pace of a reaction in both directions. The direction depends on the factors present at that particular time. It should be specified here that an enzyme by itself does not initiate a reaction but merely speeds up the rate. Thus, an enzyme *starch phosphorylase* catalyzes the hydrolysis of starch into glucose during day, but brings about the synthesis of starch during darkness. So here, it is not the enzyme that decides the direction of a reaction, but the presence or absence of light, pH etc., which do so. The enzyme merely speeds up the rate of reaction in whichever direction it is proceeding.

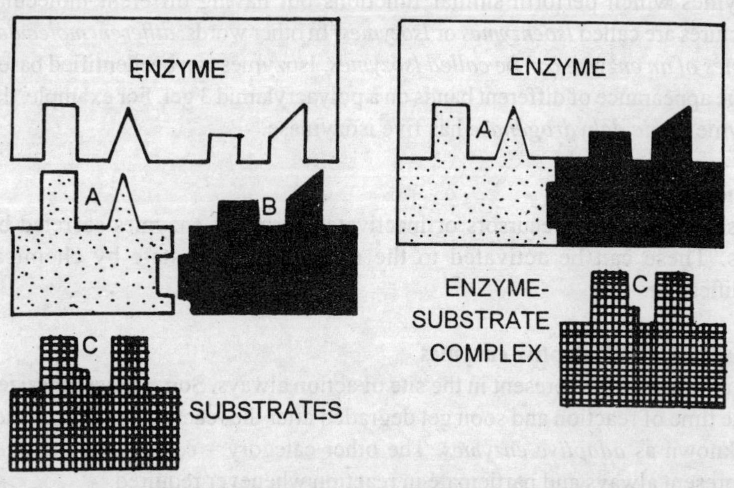

Fig. 3.1 Enzymes

Diagram to illustrate enzyme substrate specificity

3. Specificity

Enzymes are highly specific in their action. A particular enzyme catalyzes only a particular kind of a reaction. Sometimes, if a reaction involves several steps, an enzyme will be specific for each step. Enzyme specificity is of the following types.

(i) *Absolute specific*	:	Acts on only one kind of a substrate. For example, *malic dehydrogenase* acts only on malic acid.
(ii) *Absolute group specific*	:	Acts on a given organic group. For e.g., *alcoholic dehydrogenase* acts only an alcoholic group.
(iii) *Relative group specific*	:	Enzyme catalyses more than one organic group. E.g., *Trypsin* acting on ester bonds involving Arginine or lysine residues.
(iv) *Optical specificity*	:	Certain enzymes can distinguish between optical isomers in the substrate and catalyze any one only. For example *-amino acid oxidase* will not act on D - amino acids and *vice versa.*

4. Heat sensitivity

Enzymes remain active under normal temperature and are denatured under high temperatures (60 - 70^{0}C). Under dehydrated conditions however, as in seeds, they can withstand temperature up to 120^{0}C. Hence enzymes may be termed *thermolabile.*

5. pH sensitivity

Most of the enzymes are active only under a particular range of pH. Any alteration in pH would render the enzymes inactive. Enzymes are denatured by strong acids or alkali. Optimum pH for each enzyme varies depending on the nature of the buffer system (if present), absorbed ions, presence of activators/ inhibitors etc.

6. Inhibition of enzyme action

Most of the enzymes have active sites to which the substrate binds to undergo conformational changes. In some instances, it has been noticed that certain other substances which are partially indentical to the substrates bind themselves to the enzyme and inactivate it. These substances are called inhibitors and their action is called inhibition. There are three types of inhibition *viz.* **competetive inhibition, non competetive inhibition** and **uncompetetive inhibition.**

(i) Competitive inhibition

In this, the inhibitor molecule is structurally analogous with the substrate molecule. It competes with the substrate molecule to occupy the active site of the enzyme. Due to the occupation of the active site by the inhibitor molecule, the enzyme gets inactivated i.e., it cannot accept a substrate molecule (Fig). For example, *succinic dehydrogenase* is inhibited by molecules of malonic acid.

Competitive inhibition can be overcome by the addition of more molecules of substrate to the system which can now favourably compete with the inhibitor molecules for the active site of the enzyme.

(ii) Non competitive inhibition

The inhibitor molecules are not structural analogos of the substrate, hence, they do not bind to the active site of the enzyme, instead they bind themselves to the enzyme in some other area, all the same, the enzyme gets inactivated due to conformational changes. Heavy metallic ions such as Mg ++, Ag++, Pb++ etc. combine with the sulphydryl groups of the enzyme rendering it inactive.

Non competitive inhibition is irreversible. Addition of extra molecules of substrate is of no consequence since the active sites are anyway available to the substrate.

(iii) Uncompetetive inhibition

In competetive inhibition, the inhibitor is supposed to combine with forms of the enzyme but does not combine with the substrate enzyme complex. The form of the enzyme to which uncompetetive inhibitor attaches will always remain substrate non combining, in the sense it can never get back to the substrate combining form of the enzyme. As a result, enzymatic catalyses will stop. This inhibition also cannot be remedied by adding more amount of substrate. Uncompetetive inhibition is very common in multisubstrate reactions, but very rarely in single substrate reactions.

6. Colloidal condition

The enzymes are large molecules of proteins and exhibit all the colloidal properties. They cannot pass through a membrane of *collodion.* The enzymes have the capacity of absorption.

7. Redox potential value

Certain enzymatic reaction which are reversible, depend on the redox potential value for the direction of the reaction

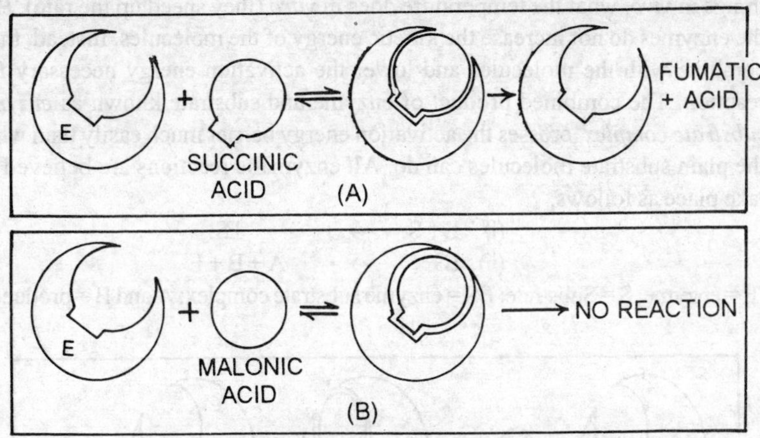

Fig. 3.2 Enzymes
Competetive inhibition of enzyme action

MECHANISM OF ENZYME ACTION

How does an enzyme catalyze, or speed up a reaction? The following explanation may be offered by considering the conditions of a biochemical reaction.

Normally, chemical reaction take place when molecules of any substance possessing on an average a higher kinetic energy change into a substance whose molecules have a lower average of kinetic energy. Although on an average all molecules of reactions (substrate) possess equal amount of kinetic energy, a few of them possess higher kinetic energy, and a few of them possess lower than the average due to collisions among molecules. Thus among the molecules of a substrate, there are *energy rich* and *energy poor* molecules. If we, for instance take a reaction where molecules of A get converted into B, at any given time only few molecules of A can get converted into B because of their higher kinetic energy. This extra energy required for the participation in the reaction is called

activation energy. The activation energy is always above the average kinetic energy of the molecules. In a non enzymatic reaction, the average kinetic energy of molecules may be raised to activation energy by increasing the temperature of the medium. This speeds up the rate of reaction. Theoretically, continuous increase of temperature should indefinitely increase the rate of reaction. But this is not possible actually because, a higher temperature beyond a certain point would kill the whole system. The enzymes in living cells do the same things in *vivo,* what the temperature does in *vitro* (they speed up the rate). But the enzymes do not increase the kinetic energy of the molecules. Instead, they combine with the molecules and lower the activation energy necessary for reaction. The combined product of enzyme and substrate known as *enzyme substrate complex,* crosses the activation energy barrier much easily than what the plain substrate molecules can do. All enzymatic reactions are believed to take place as follows.

(i) E + S → ES

(ii) ES → A + B + E

(E = enzyme; S = Substrate; ES = enzyme substrate complex; A and B = products)

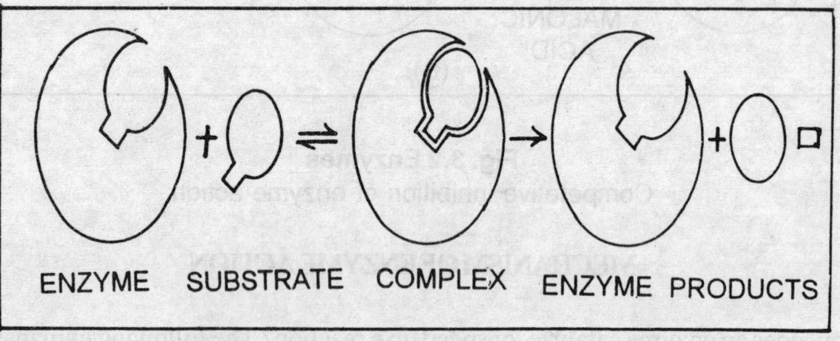

Fig. 3.3 Enzymes
Graph to demonstrate lowering of activation of energy due to enzyme catalysis

Measurement of activation energy has been done for several reactions with or without enzymes. With the participation of enzymes, it has been noticed that more number of substrate particles can participate in a reaction than what is normally possible. For example, hydrolysis of sucrose with only hydrogen ions as catalyst has an activation energy of 25,000 cal/mol., but with the addition of yeast *invertase,* it comes down to 10,000 cal/mol. Similarly, in the decomposition of H_2O_2 (enzyme *catalase*), activation energy comes down from 18,000 cal/mol to 2000 cal/mol; for caesin the figures are 20,600 cal/mol and 12,000 cal/mol without, and with enzyme (*trypsin*).

Enzyme kinetics (Michaelis - Menton equation) : Michaelis and Menton (1913), proposed a theory for enzyme action, in which they believed that the substrate and enzyme form a loose complex, and at the end of reaction the complex breaks releasing the enzyme and products as explained above. In order to explain the rate of substrate conversion, they applied the laws of kinetics which may be represented as follows.

$$E + S \xrightarrow{K_1} ES$$ (E = enzyme; S = substrate; ES = enzyme substrate complex; K_1 velocity constant of the formation of ES and K_2 velocity constant of dissociation of ES)

OR

$$E + S \xrightarrow{K_1} ES \xrightarrow{K_2} P + E$$ (K_2 = velocity constant for decomposition of ES.P = end product)

The equation can be further simplified as $\frac{K_1 + K_2}{K_1} = K_M$

K_M is known as the velocity constant, or Michaelis and Menton's constant. Its value is characteristic for each enzyme and substrate and by comparing KM values of an enzyme with different substrates, the affinity of different substrates to an enzyme can be calculated. When the velocity of the reaction is half of the maximum velocity, KM is equal to the concentration of the substrate. However, Michael is Menton equation is applicable only to single substrate reactions.

MODELS FOR ENZYME ACTION

Two models have been proposed to account for the specificity and mode of enzyme action. These are (i) the lock and key theory, and (ii) the induced fit theory.

Lock and key model

This theory originally proposed by Fischer (1898), believes that the enzyme specificity can be accounted for by assuming that the enzyme and the substrate fit into each other structurally like the two pieces of a jigsaw puzzle or the lock and key. Just as only a particular key can fit into a particular lock because of the alignment of levers, enzymes have specific active sites, to the contours of which, the substrate molecules can fit in.

No other molecule can fit in to this slot. When a reaction mixture consists of several substrates, like A,B,C,D etc., only A and B can fit into the active sites and undergo conformational change resulting in the product P (This is an example molecule, into two products, only one active site will be present on the enzyme and only one substrate A or B can fit into the active site.)

The lock and key mechanism assumes a rigid structural configuration for the enzyme molecule and does not allow even minor adjustments between the enzyme and the substrate.

(ii) Induced fit model

Many experimental findings questioned the rigid unchangeable configuration for the enzyme molecule as proposed by the lock and key model. It has been noticed that the enzyme molecule does allow for minor adjustments in its geometry to permit the substrate to fit into the active site. Secondly, it has also been observed that the substrates induce proper orientation of the catalytic groups in the enzyme in order to activate enzyme action.

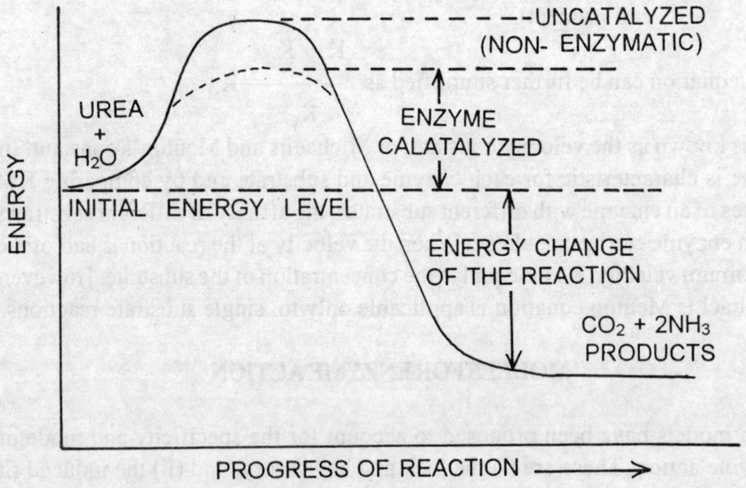

Fig. 3.4 Enzymes

Lock and Key mechanism of enzyme action

In order to make provisions for the above findings, Koshland (1959), proposed a theory called the induced fit theory. According to the induced fit model a substrate after linking itself with the enzyme alters the geometry of the enzyme protein in such a way as to achieve perfect alignment.

Allosteric regulation

In reactions which involve several sequential steps many enzymes participate, each one catalyzing a particular step. Such enzymes constitute the multienzyme complexes. In multienzyme complexes, the product of one step is a substrate for the next step. The intermediate products are transferred from one enzyme to the other. In these enzyme complexes, there is end product inhibition of all the enzymes involved. In otherwords, the end product of the pathway blocks the functioning of the first enzyme (of the complex), by binding itself to a site called the *allosteric site,* which is different from the active site. The end product which binds to the allosteric site is called the *modulator* and the enzyme itself is known as the allosteric enzyme (allosterism means another space). The name allosteric enzyme was first proposed by J. Monod, J.P. Changeux and F. Jacob.

Allosteric regulation is different from non competetive inhibition in that, it is the *end product* of the pathway which blocks the enzyme action of the first in the series of the enzyme complex. Allosteric regulation need not always be inhibition of enzyme action *(negative modulation).* In some cases, the end product may even activate the first enzyme of the series *(positive modulation).*

Mond, Wyman and Changeux (1965), proposed a theoritical model to explain the allosteric regulation. According to them, the allosteric enzymes are oligomers made up of identical units called monomers. The monomers are linked together in such a way that all of them occupy equivalent positions. Each monomer possesses one site for the binding of each molecule of either the substrate or the allosteric effector which are able to bind with the enzyme. The three dimensional configuration of each monomer is dictated by its association with other monomers. The oligomer molecule can exist in atleast two states which differ in the distribution and energy of interpromoter bond. When a transition from one state to another state occurs, there is a change in the affinity of the binding sites towards the molecule.

Double displacement (ping pong) reactions : In enzyme reactions involving a single substrate, it is well known that the substrate concentration influences the rate of the reactions. In many biological reactions however, a single reaction with the enzyme involves two or more substrates participating simultaneously. For example in the following reaction involving *hexokinase,* ATP and glucose are substrates with ADP and glucose - 6 - phosphate being the products.

hexokinase

Glucose + ATP ———————— Glucose 6 phosphate + ADP

In a situation as mentioned above, what will be the influence of the substrate concentration on the enzyme activity? In several bisubstrate reactions, it has been observed that the enzyme has a separate velocity constant (KM) for each of the substrates involved.

Bisubstrate reactions usually involve the transfer of a functional atom or group from one substrate to the other. This may proceed in one of the two following ways – (a) *Single displacement method* and (b) *Double displacement method.* In the first type, both the substrates bind to the enzyme simultaneously to form the enzyme substrate complex.

A + B + E EAB

(A - substrate EAB - Enzyme substrate complex B- substrate E - enzyme)

EAB then breaks up to form the products C and D

EAB ———— C + D + E

In the secondary category of bisubstrate reaction called the *double displacement type,* only one substrate can combine with the active site of the enzyme at any given time. After the substrate transfers its functional group to the enzyme molecule, it (substrate) leaves the enzyme as the product. The second substrate now binds to the enzyme (which has the functional group of first substrate), accepts the functional group donated by the first substrate and leaves the enzyme as product.

E + A   EA
EA   E' + C
E + B   E'B
E'B   E + D

(A and B = substrates, C and D = products, E = enzyme EA = enzyme substrate complex, E'=enzyme with the functional group of substrate A, E'B = enzyme substrate complex; enzyme has the functional group of A).
The enzyme which has the functional group of one substrate after it (substrate) has been released as the product, may be called a *partial enzyme substrate complex.*

This type of a reaction in which the substrates alternately bind and get released to the enzyme is called **ping pong reaction or double displacement reaction.**

Transamination reactions involving the enzyme *transaminase* provide a typical example of ping pong reactions. In a reaction involving glutamic acid and oxaloacetic acid, the enzyme glutamic transaminase has pyriodoxal phosphate as the prosthetic group. In the first of the reaction, glutamic acid binds to the prosthetic group of the enzyme (pyridoxal phosphate), and donates the amino group and is released as a α keto glutaric acid (product). In the second step, oxaloacetic acid binds to the pyridoxal phosphate (which has the amino group bound to it), accepts the amino group and is released as Aspartic acid (product).

```
    COOH                        COO
     |                           |            H
    CH2        E — C — H        CH2           |
     |             ||            |     +  E --- C --- H
H — C — NH3        O            CH2           |
     |                           |           NH2
    COOH                       C = O
                   Enzyme        |        Enzyme with
Glutamic acid     (Pyridoxal     |        amino group
                               COOH
           (Pyridoxal phosphate)   α ketoglutaric acid
```

```
COOH                        COOH
 |          H                |
CH2         |               CH2
 |    + E — C — H            |        +  E — C — H
C = O       |             H − C — NH3           ||
 |          NH2                                  O
COOH                        COOH   Enzyme Pyridoxal   Oxalo
acetic acid          Asparticacid   Phosphate)
```

FACTORSAFFECTINGENZYMEACTION

The following factors influence the enzyme action.

1. Enzyme concentration

At specific concentrations, and in the presence of large number of substrate molecules, the reaction will proceed at a particular speed. Any addition of substrate at this stage is not of any consequence as all the enzyme active sites are already occupied. In such a situation, increasing the concentration of the enzyme will increase the rate of the reaction. This increase stops at a point when substrate molecules become a limiting factor.

Substrate concentration

As the enzyme molecules are very large in size when compared with the substrate molecules, relatively large amount of substrate is required to engage all the enzyme molecules. At a particular concentration of the enzyme, if the substrate molecules are at a low concentration, the rate of the reaction will not proceed at the required pace since many active sites of the enzyme are unoccupied. At this stage, addition of more substrate to the system would accelerate the pace of the reaction as many enzyme molecules can be engaged, indefinite increase of substrate concentration, however would increase the pace of reaction indefinitely, as enzyme concentration would become a limiting factor then.

Temperature

Like every other bilogical reaction, enzyme catalysis is also influenced by temperature. Since enzymes are proteinaceous, they are very sensitive to temperature changes. They can function at a very narrow range of temperature since enzymes are functional even at as low a temperature as 0^0C, but the rate of the reaction will be its minimum. A rise in temperature at this stage would bring in a geometric progress in the rate of the reaction. Very soon, this reaches a peak and this temperature level is known as the optimum. Further, increase beyond the optimum temperature would actually retard the rate of reaction. A higher temperature of 50^0C and above would incapacitate the enzyme.

The Q^{10} for all enzymatic reactions is about 2.0. The higher value of Q^{10} clearly shows that temperature has a marked influence on enzymatic reactions.

Hydrogen ion concentration (pH)

Most of the enzymes are catalytically active only at a particular pH. Sometimes, they may have a range of pH at which their activity will be sustained. Any increase or decrease from this pH would render the enzymatic activity impossible. The different enzymes, however, would vary in their pH requirement. While some enzymes are active at a low pH, others prefer an alkaline pH; still others require a neutral pH for their optimum activity.

Inhibitors

The presence of certain inhibitor molecules in the system would render the enzyme inactive. Various substances are known to inhibit enzyme action. These are of 3 kinds.

1. Competitive inhibitors
2. Non - competitive inhibitors
3. End product accumulation

In many cases, if the end products of a reaction are not removed from the system they accumulate and bring about inhibition of enzyme action known as feed back inhibition. The feed back inhibition many a time may be due to allosteric effect as can be seen in reactions involving multienzyme complexes. The end product inhibition of enzyme action may also be due to a difference in the pH value of the substrate (at which enzyme is active), and of the end product.

4

BIOENERGETICS

All living cells are virtual biochemical factories where hundreds of reactions take place every second involving conversion, degradation and synthesis of chemical compounds that are vital for the maintenance of life. These reactions require energy or produce energy whic his either obtained from the breakdown of chemical compounds synthesis, or in the bonds of chemical compounds. In other words, these reactions are accompanied by energy changes. Thus, *bioenergetics may be defined as the study of energy changes that accompany biochemical reactions in living systems.* Bioenergetics may also be called biochemical thermodynamics, as the energy changes follow the laws of thermodynamics precisely.

In any biological system, the rectants have a higher energy level and they move **"energy down hill"** after the reaction. The energy thus liberated is wasted as heat either partially, or wholly. The energy that is liberated as heat cannot be put into any biological function and thus is a waste. The living systems, however, have developed means and mechanisms to trap the energy and store them in the form of energy rich chemical bonds. These 'energy currencies' readily donate the energy as and when required to do so, and help perform all the activities of living beings. One such universal energy currency is a molecule of ATP (Adenosine triphosphate). Whenever energy is required for a process, ATP is hydrolysed to ATP and the energy so released is utilized for the necessary activity. Similarly, there are a number of methods in which ATP molecules are synthesized (as during respiration).

The basic rules of bioenergetics are no different from those applicable to physico - chemical energy transformations. The only difference, perhaps, is that biomolecules involved in bioenergetics are more complex and function in a highly specialized environment. In order to understand the principles involved in bioenergetics it will be necessary first to understand the physiochemical principles involved in energy transformation reactions, and later these can be applied to biological situations.

Energy and matter

The universe is composed of only two components – *energy* and *matter* which were thought to be non interconvertible and fixed in their content. Einstein ,however, has shown that matter can be coverted into energy. What is energy? Energy may be defined as the ability or capacity of a system to carry out some work. This capacity is there in all the molecules of matter to a lesser or greater extent.

Types of energy

Basically, there are two main types of energy. These are – *potential energy* and *kinetic energy.* Potential energy is bound to the molecules. It may also be called as energy at rest i.e., not working. It can be measured in terms of how much work it is capable of doing when it is released. Potential energy is of various types – chemical, light (photic), atomic and even positional (such as that of a boulder on hill top). In biological world, potential energy source is mainly the solar radiation which is trapped, transformed and stored in the form of chemical energy during photosynthesis in greeen plants. All living beings obtain their energy requirements from the green plants directly or indirectly.

Kinetic energy is the other form of energy. It can be called as *energy in motion,* or energy released. Molecules or bodies have this energy as a result of motion. The amount of kinetic energy depends upon the mass of the body and also its velocity of motion. The following equation summaries this relationship.

$$KE = \frac{mv^2}{2} \quad \text{or} \quad KE = 1/2\ mv^2\backslash v^2\backslash v^2$$

In the above equation KE = Kinetic energy, *m* = mass of the body/molecule, and *v*= velocity of the body/molecule.

Kinetic energy of molecules also depends on one other parameter viz., temperature. The amount of molecular kinetic energy in a system is proportional to the absolute temperature of that system and at absolute zero (-273.18^0 C), kinetic energy disappears, i.e., it is zero. For every rise of 10^0C in temperature, kinetic energy increases by the ratio of 10 to the absolute temperature.

Kinetic energy of molecules may be easily demonstrated through the Browinian movement of the particles. Brownian movement, is the to and fro movement of medium. This can be clearly seen when a beam of light is made to pass through a colloidal medium. The rate of movement of particles increases with the increase in temperature. The mechanism of Brownian movement is attributed to the random motion of molecules colliding with the visible particles, and imparting to them required energy for movement.

In addition to kinetic and potential, energy may also be of two other types. Free energy and Internal energy. Free energy may be defined as the energy that is liberated or consumed in a reaction which can be used or supplied to another system. Free energy is the most important type of energy in the biological system (see later in the chapter for a detailed discussion on free energy).

Internal energy is the energy that is associated with a system based on the constitution and movement of the molecules. Internal energy is again of two types – *Internal potential energy* and *Internal kinetic energy*. The contribution of energy due to its molecular configuration is Internal potential energy, and the energy derived from the motion of its molecules is Internal Kinetic energy. The total internal energy of a system is the sum total of both the above types of energy.

In any given substance, the internal energy is fixed under a given set of conditions. The internal energy of different substances, however, varies. Internal energy is symbolically represented by E or U.

Matter

This may be defined as something that physically occupies space and of which the entire physical universe is composed of. Originally, it was thought that matter can neither be created nor destroyed, but since the epoch making work of Albert Einstein, it has been found out that two atomic reactions *viz; nuclear fusion* and *nuclear fission* can convert matter into energy. But nuclear fusion or fission requires a very high temperature range (Nuclear fission occurs in sun and nuclear fusion is seen in atomic bombs), and they do not occur in ordinary chemical reactions.

The study of matter and its energy, conversions follow the principles of thermodynamics. In order to study the principles of thermodynamics as it relates to matter, an arbitrary position of matter, a system and its surrounding has to be imagined/considered. A system may absorb energy from the environment of liberate energy into the environment. In biological situations, *a system* may be a cell and its surroundings, or an organism and its environment or it may be an ecosystem. But the principles of thermodynamics remain the same irrespective of the size and structure of matter.

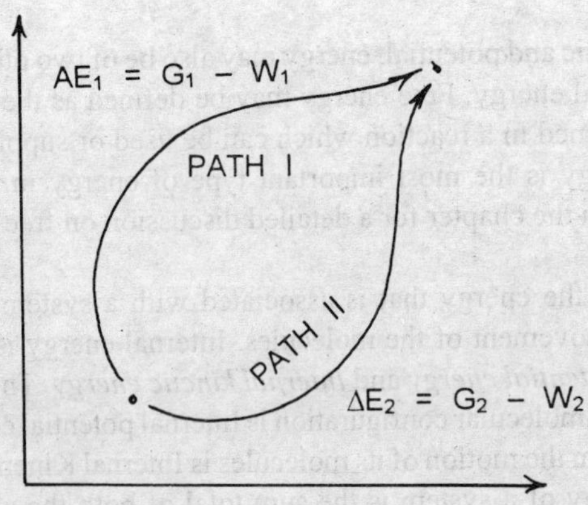

Fig.4.1 Bioenergetics
Illustration of pathways for energy conversion of *a* into *b*

Laws of Thermodynamics

It has already been pointed out that the bioenergy conversions follow the principles of thermodynamics. Hence, an understanding of the laws of thermodynamics becomes necessary before attempting to understand the role of energy in biological reactions.

It has been mentioned above that irrespective of the size and structure of matter, the laws of thermodynamics are the same. Classic thermodynamics in reality deals with reactions in the equilibrium state, and may be regarded as *equilibrium thermodynamics*. Equilibrium thermodynamics is not concerned with the pace of the reaction, instead, it is concerned with the change in the form of energy from the initial stage (before the reaction has begun), to the final stage (when the reaction has been completed). Some of the other salient features of classic thermodynamics are –

(i) The number of steps between the initial and the final are irrelevant as long as the final state is same in different systems.
(ii) Direction of the reaction can be indicated in reversible reactions.

There are two laws of thermodynamics which are universally applicable. These are –

The first law of thermodynamics

This is known as the *Law of conservation of energy.* It states that the total content of energy is fixed and it can neither be created nor destroyed. But during reactions, energy can change form. In other words, the total energy in an isolated system will neither decrease nor increase. For instance when some amount of chemical, light or electric energy disappears (apparently), it would be transformed into equal amount of heat.

Previously, many attempts were made to disprove the principle of conservation of energy by constructing a 'perpetual motion machine', which is supposed to operate without any input of energy i.e., creating energy out of nothing. But these attempts failed.

The first law of thermodynamics may be enunciated in a detailed manner by the two following statements. These are –

(i) The disappearance of energy in one form will result in its reappearance in another form in exactly the same quality.
(ii) Energy of an isolated system will remain constant although it may be transformed from one form to another.

Elaborating the first statement, it may be said that from an isolated system when heat is given out (as in exothermic reactions), the internal energy of the reacting system decreases while that of the surrounding medium increases correspondingly. Hence, the total energy of the system (energy of the system + surrounding) remains unchanged. Alternatively, when there is energy input into the system (absorption of heat), the energy of the reacting system increases while that of the surrounding decreases. In this case also, the total energy remains unchanged.

As an illustration to elaborate the second statement, the burning of coal in a steam engine may be cited. Coal has a lot of internal energy. When coal is burnt, a lot of energy is released as heat, which is converted into mechanical energy of the engine i.e., it is allowed to boil water and release steam to rotate the wheels. In this, the chemical energy of coal gets transformed into mechanical energy of the engine.

Energy exchanges in chemical systems are studied with the help of a 'bomb calorimeter'. In the working principle, a calorimeter is nothing but a slightly modified thermos or Dewar flask. Basically, it has an insulation mechanism to isolate a system with internal energy as in a thermos flask.

A bomb calorimeter consists of an outer jacket (made up of insulating material not to allow heat to escape). This jacket surrounds a chamber which is filled with water. The centre of the calorimeter has a central chamber (combustion chamber) containing a sample which releases heat. The heat released by the sample increases the temperature of the water and its temperature is noted with the help of a thermometer. Since a calorimeter may not have a 100% insulation, corrections must be made for possible thermal leaks. With the help of this, it will be possible to show that the rise in temperature of water corresponds to the quantity of heat released by the sample.

Calculations of total energy (as per the first law)

Calculations of total energy in a reacting system can be made by finding out the quantities of the following :

(i) The work done
(ii) The heat that is exchanged, and
(iii) The energy stored in the system

Mathematically the above may be expressed as follows.

$$\Delta E = q - w$$

Where ΔE is change in the internal energy of the system, q is the heat absorbed by the system and w is the work done.

The values of q and w may be positive or negative, depending on the nature of reaction. If heat is absorbed, q has a positive value. If heat is given out, q has negative value. In the same way if the system performs work, w will be positive and if the medium performs work on the system, w will be negative.

The change in total energy does not depend whether q and w are negative or positive. For instance if molecules of *b, the change in energy is* ΔE_l, and the other case ΔE_2 then $\Delta E_1 = \Delta E_2$ The energy change her may also be represented as –

$$\Delta E = Ea - Eb$$

Where Ea and Eb are energy levels of molecules of *a* and *b* respectively. The first law of thermodynamics needed some modification when Albert Einstein discovered that matter is convertible into energy. Accordingly, it is now agreed that matter (mass) can be converted into energy according to the following equation.

$$E = mc^2$$

Where E = energy; m = mass and c is velocity of light in cm/sec.

The reverse of this system i.e., conversion of energy into mass is also possible. Based on these observations, the first law of thermodynamics may be restated as *matter and energy can neither be created nor destroyed but can be inter converted.*

The interconversion of matter and energy however, does not occur in biological systems.

Second law of Thermodynamics

As per the first law of thermodynamics, the total energy in a reaction remains unchanged, but the transformation of energy from one to another proceeds in such a fashion that the capacity of the total energy to do work decreases. This is because all natural processes have a tendency to move towards equilibrium. The second law is actually related to the equilibrium of the process. Stated in very simple terms, it means during every energy change the capacity of total energy to do work decreases.

The second law of thermodynamics states that spontaneous reactions have a direction for the flow of energy. Once a reaction has reached an equilibrium, the reactants cannot return spontaneously to the initial stage.

Potential energy of molecules is often referred to as energy with a high degree of orderliness, whereas the kinetic energy is said to be random or disorderly. When once potential energy changes into kinetic energy, its randomness or disorder increases. This increase in randomness is called entropy. The stage of entropy is energy in that state where energy is incapable of doing any work because of its disorderliness. In otherwords when energy is transformed from one state to another state, the entropy goes on increasing untill it reaches an equilibrium. At equilibrium, entropy is maximum.

In nature, randomness or disorder is more a rule than exception. This is the reason why orderly arrangement which is special, tends to always move towards randomness. Entropy can be zero if a system is fully orderly. For instance, at absolute zero, the entropy of a crystalline substance is zero because there is no randomness.

The second law also stipulates that in any transformation of one type of potential energy into another type, the efficiency of the process is not perfect because some quantity of potential energy gets transformed into kinetic energy. Although theoritically, total kinetic energy can be transformed into potential energy, practically it is not possible because of inefficiency of potential energy conversions. Some transformation does occur as in the case of steam turbine or internal combusion engine, but in these instances only a small fraction of the kinetic energy becomes potential mechanical energy.

Ultimately, all potential energy of the system (whether it be a molecule, body, or even the entire universe) gets converted into kinetic energy or heat and in course of time (in finite), the entire potential energy may get exhausted i.e. Total entropy.

The fundamental principles of the second law of thermodynamics also operates in a food chain, in an ecosystem where starting from producers with a high degree of order (more number of individuals), succeedingly there will be less and less consumersin number of the primary, secondary and tertiary variety respectively. This is due to the fact that the whole of the potential energy of a producer cannot be transformed into the potential energy of a consumer. This is because of the intervening kinetic energy which introduces some amount of randomness (entropy).

Operation of the principles of second law of thermodynamics can be seen in our day to day life. For example water runs downhill and not uphill and when it has flown down its capacity to work gets decreased.

Green plants in this respect can be regarded as the most efficient mechanism to store energy for their own use and and for the use of other life forms. They trap the radiant energy and store an appreciable amount of it in the form of chemical energy. It is this potential energy which forms the fulcrum on which revolve all forms of life.

The concept of free energy

One of, the most important thermodynamic concepts that is of great relevance to biological systems is that of **free energy.** It is the energy that is liberated or consumed in a reaction and can be used or supplied to a system. It is denoted by the letter *G.* It may be regarded as a measure of the potential energy of a substance but it cannot be measured directly. However, change in free energy content that accompanies a reaction can be measured. This change is denoted by ▲G, may be defined as that quantity of total energy change in a system that is available to do the work. Change in free energy (▲G) of a reaction is calculated by considering the difference between the sum of free energies of the reactants and products. This may be represented as follows.

A ——— *B*

In the above, *A* is reactant and *B* is in the product. Here, free energy is the maximum energy made available, as *A* is getting converted to *B*. If the free energy content of the product B *(Gb)* is less than the free energy content of the reactant A *(Ga),* then ▲G will be negative (Here, reactant has more energy than the product). This may be shown as follows :

▲G = *Gb-Ga*

Here, *Ga* is greater than *Gb,* hence ▲G is negative. There is decrease in energy level as A gets converted to *B*. Similarly, when the reaction is reversible, *B* gets converted to *A*. In this reaction the level of free energy is increased (as B has less free energy).

In this case ▲G will be positive. Based on the fact whether ▲G is positive or negative (whether it gets reduced or increased), reactions are categorized into two types **endergonic** and **exergonic.**

Exergonic reactions

These reactions do not require the input of external energy. Normally, reactants have more energy than products. Products are *downhill* in energy level while reactants are *uphill* in energy level. Hence, no energy input from the medium is required to carry out the reaction. The reaction is driven by the inherent potential energy of the reactant molecules. Hence, the reaction can take place without any external energy supply. During exergonic reactions energy is released.

Endergonic reactions

Here, the reactant molecules have less free energy than the products. That is, product is *uphill,* and reactant *downhill.* Naturally, the direction of the reaction cannot be from down hill to uphill. The reactant molecules cannot negotiate on their own the uphill migration. Hence, energy from outside has to be supplied to the reactants to make them participate in the process. Here ▲G is positive *(GB–GA* and *GB* is greater than *GA,* where *GB* is free energy of reactant, and GA is free energy of the product).

The difference between endergonic and exergonic (energy requiring and energy releasing) processes can be explained with an example. Suppose a huge boulder is on top of hill, no extra effort is required to push it down hill; just a shove would do it. It has greater potential energy because of its position on hill top. When it reaches foot hill, its free energy is less because of its different position (location). This can be called exergonic process. Suppose the same boulder has to be now pushed back (from foot hill) uphill it requires an enormous effort (input of energy), which should come from outside and comparable to input of energy into a system. The boulder that is pushed uphill will again have more amount of free energy than the boulder at foot hill. This process is called endergonic, as there is energy requirement.

Photosynthesis and respiration serve as the best examples of endergonic and exergonic processes. For instant the overall formula of the processes.

1. $C_6H_{12}O_6$+6O_2+6H_2O ——— 6CO_2+12H_2O+ energy (respiration)
2. 6CO_2 + 12h_2O $\xrightarrow{\text{light energy}}$ $C_6 12_{12}O_6$ + 6O_2 6H_2O (photosynthesis)

In the above equation for respiration, reactant (glucose) has more free energy than products. Hence, along with products energy also will be released. Reaction takes place without any energy input into the system as the reactants are *energy uphill,* and products *energy downhill.* The reaction is said to exergonic.

Equation for photosynthesis on the other hand shows that product viz., glucose has more free energy than the reactants viz CO_2 and H_2O. Hence, to drive the reaction *uphill,* there is requirement of energy input into the system which is provided by light energy. The reaction here is said to be endergonic.

Calculation of standard free energy change

In an exergonic reaction where ▲ G is negative, the rate of reaction however, is determined by the surrounding conditions like temperature (heat) etc. For instance suppose 10 molecules of reactants have to become products, not all the 10 molecules may participate in the process as they may have lost some energy because of collision among molecules. Hence, only some molecules which have the required high energy participate in a reaction. This is known as *activation energy.* Activation energy for a process is the higher potential energy that is required to start a reaction. Acvitation energy may sometimes be increased depending on the environment, like a medium having higher temperature. This means then the reactant molecules have varying speeds of the reaction even in exergonic reactions depending on the medium. For example, glucose can be oxidized to CO_2 and H_2O in different surroundings. In a bomb calorimeter, oxidation takes place in a few seconds; in living organisms it takes minutes to several hours. Similarly, even in the presence of O_2 (oxidizing agent), glucose can be kept in a bottle for years without large scale oxidation. In a reaction where A and B are reactant and product respectively.

$$\Delta G = \Delta G^0 + RT\, B/A \Delta$$

Here ΔG^0 is standard free energy. R is the gas constant and T is absolute temperature in Kelvins. This equation shows that when conditions are ideal ΔG and ΔG^0 are equal (ΔG = free energy and ΔG^0 is standard free energy). *Standard free energy change may be defined as the free energy change of chemical reaction when reactants and products are in their standard state.* In otherwords, standard free energy change of a substance is that value which it can ideally release when all conditions are optimum. Standard free energy change of a process can be used to find out whether a reaction is proceeding along the ideal path of not. If there is any difference between the standard free energy change and the actual free energy change that is going on in a reaction, it indicates that the conditions are not ideal necessiating corrective measures.

To summarize, the standard free energy change can be calculated as follows :

$$A + B \_\_\_\_\_\_ C + D \quad (1)$$

Where A, B, C and D are molecules of substances participating in a reaction. The free energy change ΔG, at constant temperature and pressure is given by

$$\Delta G = \Delta G^0 + RT \text{ in } \frac{C+D}{A+B} \quad (2)$$

In the above, ΔG = is free energy change; ΔG^0 = standard free energy change; R=gas constant and T= absolute temperature.

In the equation (2), above ΔG^0 has a definite (fixed) value characteristic for any given chemical reaction. Thus, we can calculate ΔG at any given temperature if ΔG^0 is known in addition to concentration of products and reactants.

If the reaction at equation (1) is allowed to proceed towards equilibrium, as the free energy decreases, the reaction will be able to carry out work at constant temperature and pressure. When equilibrium is reached, ΔG will be at its minimum or zero, and the system halts as it cannot do any more work. The equation at equilibrium is

$$O = \blacktriangle G^0 + RT \, In \, \frac{C+D}{A=B} \quad (3)$$

If we rearrange the above equation

$$\blacktriangle G^0 = - RT \, In \, \frac{C+D}{A+B} \quad (4)$$

Since the equilibrium constant K for the reaction would be

$$K = \frac{C+D}{A+B} \quad (5)$$

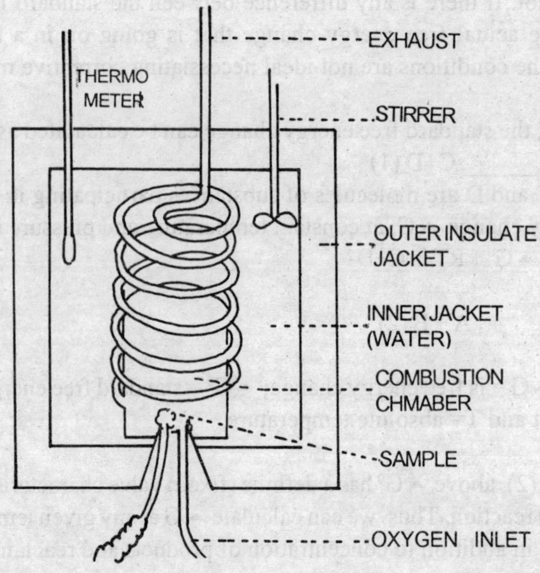

Fig. 4.2 Bioenergetics

A bomb calorimeter for determining heat of combustion of compounds

K can be substitued in the equation (4) to state as follows

$$G^0 = -RT \text{ in } K \text{ or } G^0 = 2.303\, RT \log K$$

The above equation shows that if we can calculate the equilibrium constant for any reaction, the standard free energy change can also be calculated. In the same way, if ▲ G^0 is known, ▲ G is calculated either in calories/mole or joules/mole depending on whether gas constant is given in calories (R=1.98 mol^{-1} K-1) or joules (R=8.31 J mol^{-1} k-1)

As a conclusion to the discussion, the following summarized statements may be made about standard free energy change

(a) "The standard free energy change of a given chemical reaction is the difference between the sum of the free energies of the products and the sum of the free energies of the reactants, each reactant and product being present in its standard state" (Lehninger, 1970). The standard condition on the various components in a reaction system are –

   (i) Cencentration of 1.0m
   (ii) Temperature 25°C or 298°k
   (iii) Pressure 1.0 atm

Each compound has a particular architectural pattern that decides its inherent free energy. Hence, the standard free energy ▲ G^0 of a chemical reaction may be expressed as follows

$$\blacktriangle G^0 = \underset{(product)}{EG^0} - \underset{(reactant)}{EG^0}$$

(b) The standard free energy change may also be expressed in terms of the equation A+B______C+D. Here, ▲ G^0 is that quantum of free energy absorbed or lost per mole when molecules of A and B get converted into molecules of C and D, when the conditions of all the components are standard. The following table gives the free energy changes of some chemical reactions.

Standard free energy changes at pH7 and at 25°C of some chemical reactions
(Adopted from Arthur Giese, 1976)

Oxidation	**▲ F**
Glucose + $6O_2$ ----- $6CO_2$ + $6H_2O$	-6,86,000
Lactic acid + $3O_2$ ----- $3CO_2$ + $3H_2O$	-3,26,000
Palmitic acid + $23O_2$ ----- $16CO_2$ + $16H_2O$	-2,338,000
Hydrolysis	
Sucrose + H_2O ----- glucose + fructose	-5,500
Glucose 6 phosphate + H_2O ----- glucose + H_3PO_4	3,300
Glycyglocine + H_2O ----- 2 glycine	-4,600

Rearrangement

Glucose 1, Phosphate ----- glucose 6. Phosphate	-1,745
Fructose 6, Phosphate ----- glucose 6. Phosphate	-400

Ionization

$CH_3COOH + H_2O$ ----- $H_3O_3+ + CH_3COO$ -	+6,310

Elimination

Malate ----- fumarate + H_2O	+750

A sample calculation of ▲G⁰

In order to elucidate the calculation of ΔG^0 from the data obtained from equilibrium constant, the reaction in which the enzyme phosphoglucomutase can be considered. This enzyme catalyses the reversible reaction of glucose 1. Phosphate to glucose 6. Phosphate. We can take 0.020m glucose 1. Phosphate add the enzyme allowing forward direction change or take 0.020m glucose 6. Phosphate and allow the reaction to go in reverse direction. In either case, at equilibrium the medium contains 0.001m glucose. 1. Phosphate and 0.019m glucose 6. Phosphate at 25°C and at photosynthesis 7.0. The equilibrium constant can be calculated as follows

$$k = \frac{\text{glucose 6. Phosphate}}{\text{glucose 1. Phosphate}} = \frac{0.019}{0.001} = 19$$

From the value of k ΔG^0 can be calculated as follows

$\Delta G^0 = RT \text{ in } K$

$= - 1.987 \times 298 \text{ In } 19$

$= - 1.987 \times 298 \times 2.303 \log 19$

$= - 1.745 \text{ cal mol}^{-1} \text{ or } - 7301 \text{ KJ mol}^{-1}$

Energy rich compounds

Living systems must get free energy from their environment in order to conduct life processes. Autotropic organisms obtain energy from sun light and their metabolism is linked to the exergonic process (respiration) to obtain energy. On the other hand, heterotrophic organisms couple their metabolism to the breakdown of complex to say that an endergonic process is always linked to an exergonic process.

In the living systems, a number of organic compounds are present which are energy rich (high energy compounds) because they undergo a large decrease in free energy change during their breakdown. These are also called energy currencies because they readily give out energy by undergoing hydrolysis. Some of these compounds are –

1. Pyrophosphate compounds
2. Acyl phosphates
3. Enolic phosphates
4. Thiol esters
5. Guanidine phosphates

Pyrophosphate compounds

Among the pyrophosphate compounds, the most important ones are ATP (Adenosine triphosphate) and ADP (Adenosine diphosphate), from the point of view of energy changes.

ATP

Historically, ATP was first discovered by C.F. Fiske and Y. Subbarow, in USA, and by K. Lohmann in Germany in 1929. After its discovery, it was thought that ATP was primarily involved in only muscle contraction reactions. Subsequent findings of Otto Warburg and Otto Meyerhof, indicated that ATP is generated from ADP during the anaerobic breakdown of glucose to lactic acid. Later, Kalckar of Denmark, and Belister of Russia, showed the generation of ATP (from ADP) during aerobic respiration also. Hydrolysis of ATP to ADP was demonstrated by Engelhardt and Lyubjmova during energy requiring reactions.

Finally, it was Fritz Lipman in 1941, who organized all the known facts of ATP into a general hypothesis concerning energy transformation reaction in living systems. Lipman opined that – ATP functions in a cyclic manner as a carrier ATP is used for various cellular activities. When fuel molecules undergo degradation, ATP is generated by the coupled phosphorylation of ADP. The ATP so generated can donate its third phosphate group for energy requiring processes such as biosynthesis of macromolecules, transport of substances across the membrane against the concentration gradient etc. As the energy from ATP is released, it undergoes cleavage to form ADP. ADP will be rephosphorylated to ADP through the energy obtained by degradation of fuel molecules thus completing the energy cycle. The terminal (third) phosphate group is the key substance that undergoes continuous cleaving and regeneration.

Fig. 4.3 Bioenergetics
Space filling model of ATP

Structure of ATP, ADP and AMP

The three compounds are basically built up on the structure of ribose of deoxy ribose nucleoside. Coupled with a Phosphate group, these occur in three states – 51 monophosphate, 51 diphosphate and 51 triphosphate. Here, 51 indicates that the phosphate group(s) is attached to the 5th carbon position.

For Adenosine nucleoside also, there are three types. These are –

(a) Adenosine monophosphate (AMP)
(b) Adenosine diphosphate (ADP)
(c) Adenosine triphosphate (ATP)

The phosphate groups of these are designated as α, β, γ. AMP has only phosphate group, ADP has α, and β, while ATP has α, β, and γ phosphate groups. The γ phosphate group is energy rich and undergoes cleaving and regeneration.

The structure of ATP was first deduced by Lohmann (1930), and subsequently confirmed through total chemical synthesis by Alexander Todd *et al,* (1948). The three compounds (ATP, ADP and AMP), are of universal occurrence in all living beings.
The sun total of these three is constant (between 2 to 10mm) in a given species. In actively dividing cells, the amount of ATP exceeds those of ADP and AMP.

At pH 7.0 in aqueous solution both, ATP and ADP are highly charged ions. ATP has four anions while ADP has three protons. Three of the four protons of ATP get fully ionized at pH 7.0 while the fourth one will be dissociated about 75% as it has a pH of 6.95. The high concentration of closely placed negative charge around the triphosphate group in ATP has a significant bearing on its ability to carry out energy transformations.

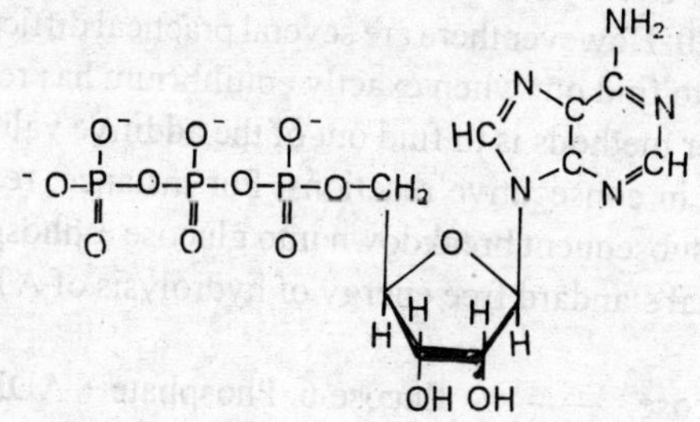

Fig. 4.4 Bioenergetics
Molecular structure of ATP

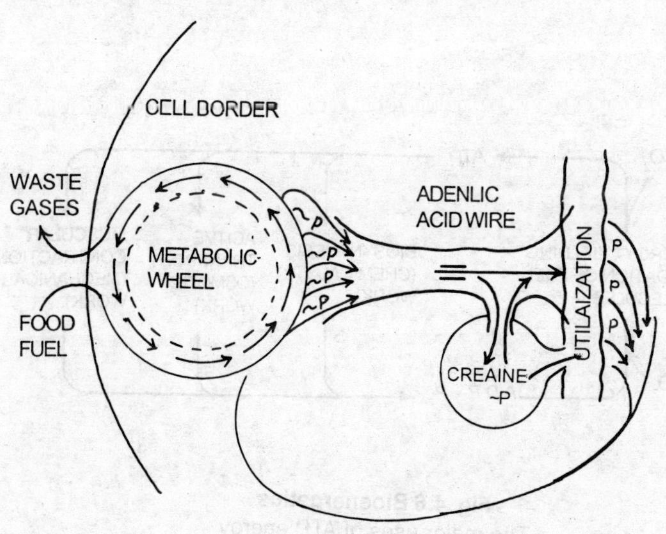

Fig. 4.5 Bioenergetics
The ATP cycle as depicted by Lipmann (1941)

Hydrolysis of ATP and Standard free energy

Theoretically, ▲G^0 for ATP can be calculated by determining the equilibrium constant at pH 7.0. However there are several practical difficulties in this method as it is not easy to find out when exactly equilibrium has reached.

One of the easier methods is to find out of the additive values ▲G^0 when ATP participants are in consecutive reactions. For instance, reaction of ATP with glucose and its subsequent breakdown into glucose + phosphate will help us to calculate the total standard free energy of hydrolysis of ATP.

1. ATP + glucose ———— flucose 6. Phosphate + ADP
   Here k = 661 therefore ▲G^0 = 4.0 cal mol$^-$
2. Here k = 171 ▲G^0 = 3.3 k cal/mol

Adding 1 and 2 of the above (-4+3.3) we get - 7.3 k cal/mol as the total standard free energy available during the hydrolysisn of ATP.

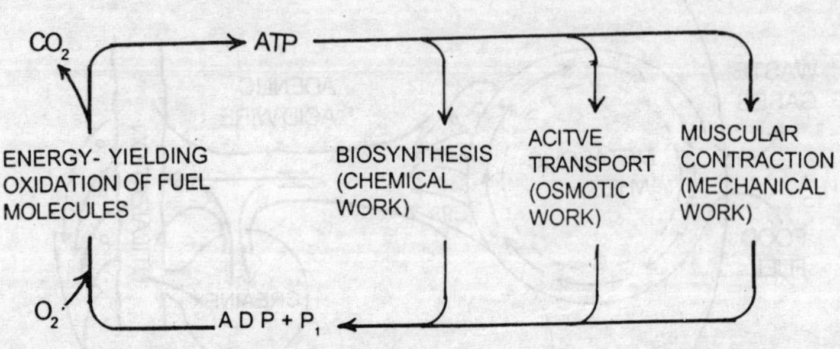

Fig. 4.6 Bioenergetics

The major uses of ATP energy

The second phosphate group of ADP also has a similar value (-7.3 k cal mol.), while the single phosphate group of AMP has G^0 value of - 3.40 k cal/ mol. This is due to the fact that the first phosphate group is linked to the nucleoside with an ester linkage, while the other two are anhydride linkages. It is well known that anhydride linkages have a great G^0 than ester linkages.

In the above discussion on energy rich pyrophosphates, attention is given to only ATP and ADP, as these participate in almost all energy transformation reactions, hence called universal energy currency. However, there are other energy rich pyrophosphate compounds such as GTP, GDP, CTP, CDP, UTP, UDP, dATP, dGTP,dTTP and dCTP (These correspond to the nucleotides of RNA and DNA). These, however, participate only in specific reactions. Thus, UTP is used in the biosynthesis of carbohydrates - polysaccharides); GTP is used in protein synthesis and CTP is utilized in lipid metabolism. These three along with ATP are involved in RNA synthesis, while dATP, dGTP, dCTP and dTTP are involved in DNA synthesis.

Another pyrophosphate compound that is energy rich is cyclic AMP (CAMP). Eventhough on hydrolysis, CAMP has a large negative free energy, it is not known to function due its high negative energy and unstable anhydride ring, instead, it acts as an allosteric effector and second messenger. Cyclic AMP is known to stimulate inactive enzymes (phosphorylase) into the active state.

2. Acyl phosphates

1.3 diphosphogylceric acid is an example of acyl phosphate. Its standard free energy on hydrolysis is 11.8 k cal/mol. The strain of the C – 0 bond is responsible for a very high ▲ G^0 in this kind of compounds. High energy is required (during synthesis) in these bonds to overcome the repulsive changes of carbon and phosphorous atoms and consequently on hydrolysis a large amount of energy is released.

3. Enolic phosphate

An example of enolic phosphate is phosphoenol pyruvic and PEP. This is formed during the glycolysis when glucose gets converted intp pyruvate. On hydrolysis, it has a standard free energy of - 14800 cal/mole at pH 7.0

The large negative ▲ G^0 observed on hydrolysis of this compound, is due to stabilization of enolic form of pyruvic acid in PEP by the phosphate ester group. On hydrolysis, the unstable enol will instantly isomerize into the much more stable keto structure. It has been in ▲ G of about 8000 cal/mol thus bringing the total upto - 14800 cal/mole. This tautomerization of PEP makes it one of the most 'energy rich' compounds in biological systems.

4. Thiol esters

An example of thio ester is Acetyl CoA. This is also an energy rich compound which can be utilized to generate ATP from ADP. The ▲ G^0 of this compound on hydrolysis is approximately -. 75v00 cal/mole. The higher negative value for ▲ G^0 is due to the fact that thioesters do not exhibit the resonance forms that is exhibited by ordinary O_2 atom. The sulphur in thio esters do not readily release the electrons for double bond formation and hence, resonance *forms possible* for oxygen ester. Usually, oxygen ester atom is knocked to the carboxyl acid carbon by means of a double bond. As the thio ester linkage is unstable (to S), the ▲ G is higher on hydrolysis.

5. Guanidine compounds

These represent a fourth type of energy rich compounds. A typical example of guanidine compound is guanidine phosphate that plays an important role in energy transfer and storage. Also known as *phosphagous,* these compounds are formed by the phosophorylation of creatine to arginine with ATP in the presence of appropriate enzyme. Hydrolysis of these compounds has a standard free energy of – 1–300 cal/mole i.e., about – 3000 more than that of ATP.

The guanidine phosphates are not inherently less stable because of bond strain as in ATP and ADP. There are no ionization or tautomerization processes which account for greater stability of reactants than their products.

In addition to the above mentioned 5 compounds which are energy rich, there are other compounds'. Some of these are, glucose 6. Phosphate, fructose 6. Phosphate etc. These have a standard free energy less than that of ATP (i.e., less than - 7300 cal/mole). The table below gives the standard free energy (on hydrolysis) of some important compounds.

Standard Free Energy on Hydrolysis of some important metabolities

	▲G' at pH 7.0 (Cal/mole
Phospheoneolypyruvate	-14,8000
Cyclic - AMP	-12,800
1,3-Disphosphyglycerate	-11,800
Phosphocreatine	-10,300
Acetyl phosphate	-10,100
S - adenosylmethionine	-10,000
Pyrophosphate	-8,000
Acetly - CoA	-7,500
ATP to ADP and Pi	-7,300
ATP to AMP and pyrophosphate	-8,600
ADP	-6,500
UDP - glucose to UDP and glucose	8,000
Glucose - 1 - Phosphate	-5,000
Fructose - 7 - Phosphate	-3,800
Glucose - 6 - Phosphate	-3,300
sn - Glycerol - 3 - Phosphate	-2,200

Coupling reactions

In biological systems, ATP is the source of energy for most of the endergonic reactions. ATP itself is synthesized in exergonic reactions by the process of phosphorylation. The endergonic and exergonic reactions are always coupled so that the energy synthesized from one (exergonic) is used to drive the other (endergonic). The net effect of this coupling is the transfer of kinetic energy from one system to the other. This coupling is mediated by a number of specific enzymes. Coupling reactions or energy coupling for efficient energy transfer is a common phenomenon. Let us consider an example in energy coupling.

The conversion of phosphoenol pyruvate to pyruvate is an exergonic process (energy releasing).

The liberated energy is used for the formation of ATP from ADP. This ATP can be hydrolysed to provide energy for other reactions. For instance, formation of glucose 6. Phosphate from glucose and phosphoric acid requires energy to the tune of – 3k cal/mole. Phosphorylation of ADP to ATP however, has free energy of –4k cal/mole. Participation of ATP in this reaction provides energy for the formation of glucose-phosphate bonds and the extra energy (of about -1k cal/mole) is liberated into the system (waste). The above reaction may be shown as follows :

1. Glucose + Pi ———— glucose 6, Phosphate + H_2O
(This is an endergonic reaction requiring energy of about –3k cal/mole)
2. ATP ———— ADP + Pi

(This is an exergonic reaction releasing energy of about –4 k cal/mole)
Here reaction 1 cannot take place unless there is energy input. In order to do this reaction 1 gets coupled with 2 mediated by the enzyme hexokinase as follows :

Glucose + ATP hexokinase glucose 6. Phosphate + ADP

Energy coupling reactions produce complex compounds apparently against the second law of thermodynamics. But it has to be noted that the algebraic sum of all the components of coupled reactions is negative i.e., the sum of energy liberated is more than the sum of energy consumed (-3 k cal/mole is used while the energy liberated -4k cal/mole). In biological systems, hydrolysis of energy rich compounds is always associated with an endergonic reaction in order not to waste the energy. This requires effective energy management which is provided.

Redox potential

The ability of any particular atom, or molecule, or ion to eject an electron (getting oxidized), is measured by its oxidation – reduction potential (ability to lose an electron - oxidation, and also receive an electron - reduction). This is known as redox potential and may be defined as *"The quantitiative measure of the affinity of a compound to lose or gain electrons"*.

During oxidative phosphorylation in mitochondria ATP is produced when electron are tossed from one acceptor to another. This flow occurs according to the relative affinities of the substrate to gain or lose electrons. This is accompanied by the alternate oxidation - reduction of the system. A compound losing an electron is oxidized, while the one accepting an electron is reduced. For instance, ferric ion is oxidized to ferrous ion by losing an electron, and when ferrous ion accepts an electron, it gets reduced to ferric ion.

Fe^{+++} oxidation Fe^{++} + e
————
reduction

A redox system can be compared to a dry electric cell. In the cell, the electrons are then transferred through a wire from one electrode to another, generating electric current. In such a system the capacity to lose or gain electrons is called the electrode potential. This can be measured through a standard hydrogen electrode. The electrode potential of a reducing oxidizing system also can be measured is a similar way. The electrode potential is here called the redox potential.

5

PHOTOSYNTHESIS

Ever since its origin, sometime in the remote past, life has been able to maintain itself in myriads of forms mainly because of its capacity of reproduction. An organism, however, has to exist,maintain itself in the healthy condition and only then can it reproduce. The existence of all forms of life on this planet has been made possible mainly by means of energy conversions. Organisms trap energy from environment which they utilize for their activities, and after a series of interconversions it (energy) is returned to the environment.

The phenomenon of energy input from the abiological environment to the biological system is always through the plants – green plants. Green plants thus, can be regarded as the basic fulcrum on which revolve all forms of life. In fact it is impossible to imagine what course the evolution of life would have taken but for the green plants. All forms of life are dependent on green plants (directly or indirectly) for their sustenance.

This unique phenomenon of energy input into the biological system is known as photosynthesis. The energy input has to be mediated only through the green plants, for they alone are capable of absorbing the energy from solar radiation.

Source of energy

There are only two sources of energy on this planet. One is, atomic energy and the other is solar energy. Of these two–atomic energy is not easily negotiable and available. Solar energy, however, is easily available in large quantities and living organisms have developed in the course of evolution mechanisms to trap and utilize this unique source of energy.

All forms of life, however, cannot directly harness solar energy. The capacity to trap solar energy is unique only for green plants and some bacteria. Green plants possess the green pigment viz., chlorophyll which can capture, transform, translocate and store energy which will be readily available for all activities of life.

Photosynthesis may thus be explained as the process in which the light energy will be converted into chemical energy. Except for green plants and some bacteria, all other organisms cannot directly utilize the solar energy; hence they are dependent on green plants for their nutritional requirements. Green plants and some bacteria are called autotrophic (they can prepare their own food), while all other organisms are called heterotrophic (they cannot prepare their own food). There is however one group of exceptional organisms which are autotrophic in spite of being non-green. These are the Chemosynthetic bacteria, which obtain energy by the oxidation of inorganic substrates such as ferrous ions and sulphur dissolved from the earth's crust of H_2S released from volcanic action. But considering the overall energy budget of lifeforms, the energy produced by chemosynthesis is very meagre and is of least significance quantitatively.

Definition

Photosynthesis (Photo = Light), may be defined as the process in which CO_2 and H_2O are put together (by green plants) into an organic molecule (sugar) utilizing the solar energy.

$$CO_2 + H_2O \xrightarrow[\textit{Green Plants}]{\textit{Light}} \text{Sugar (Carbohydrates)}$$

Photosynthesis may also be defined as the reduction of CO_2 to glucose ($C_2H_{12}O_6$) with the **H** necessary for reduction coming from water.

Photosynthesis involves both oxidation as well as reduction ultimately in synthesize energy rich compounds. Water is oxidized (electrons are removed), and the **H** released from water is used to reduce CO_2 to carbohydrates. Oxidation of water and the units of light energy (absorbed) produce energy intermediary compounds called ATP (Adenosine Triphosphate), and reduced coenzymes called NADPH + $H_{+|}$ (Nicotinamide adenosine diphosphate). These two compounds help in the fixation (reduction) of carbon dioxide into carbohydrates. These carbohydrates in turn constitute the starting point for the synthesis of various other metabolities necessary for the survival of the organism.

The definition and brief explanation of photosynthesis given above sums up the significance of the process. A few other definitions (different ony in the usage of the words) of photosynthesis are given below.

1. "Biosynthesis of simple carbohydrates from CO_2 and H_2O in the presence of sunight inside chlorophyll containing cells".
2. "Production of carbon containing compounds from CO_2 and hydrogen down by illuminated green cells".
3. "Photosynthesis is a redox process"
4. "Photosynthesis is a sensitized photochemical oxidation and reduction mechanism between water (H_2 donor) and CO_2 taking place at biological temperature".
5. "Photosynthesis is the formation of carbon containing compounds from CO_2 and water by illuminated green cells, water and oxygen being the by products".
6. Photosynthesis deals with the capturing and transforming of light energy into chemical energy".

The overall reaction of photosynthesis may be represented as follows

$$6CO_2 + 6H_2O \xrightarrow[\text{Chlorophyll}]{\text{Light}} C_6H_{12}O_6 + 6O_2$$

Oxygen which is released as a byproduct originates from water (Ruben and Kamen, 1941). In order to account for this, the equation should be changed as follows –

$$6CO_2 + 12H_2O \xrightarrow[\text{Chlorophyll}]{\text{Light}} C_6H_{12}O_6 + 6O_2 + 6H_2O$$

MAGNITUDE OF PHOTOSYNTHESIS

As has already been pointed out, photosynthesis is the most important biological process that supports all forms of life. Either directly or indirectly, all forms consume photosynthetic energy. From the point of view of human welfare, the contribution of photosynthesis is immense. Coal, gas oil etc., represent the photosynthetic products of the plants belonging to early geological periods.

The significanse of photosynthesis as a vital process contributing to the survival of life can be understood when we take into consideration the overall; rate of photosynthesis.

According to an estimate, the overall area under green vegetation on this earth is approximately 510 million sq.km. Of this, oceans alone account fo r361 million sq.km.

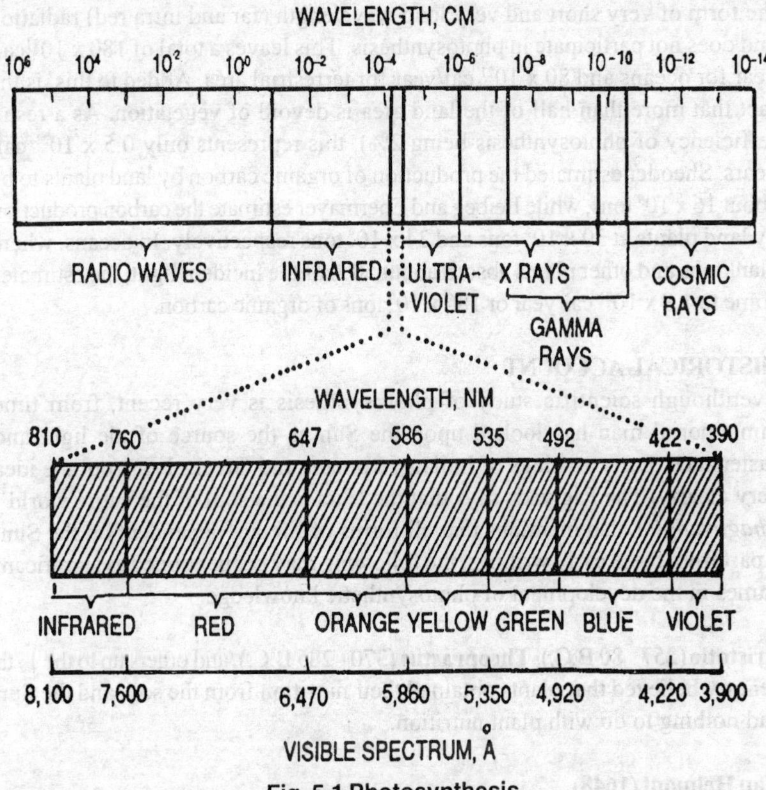

Fig. 5.1 Photosynthesis
Electromagnetic Spectrum

Regarding the availability of light, solar radiation which is emitted from the sun in the form of electromagnetic radiation reaches the outer atmosphere. In terms of energy, the outer atmosphere receives about 1300 x 10^{21} cal/year. Much of this is shielded by the atmosphere and the surface of the earth receives light equivalent to 650 x 10^{21} cal/year. Considering the distribution of water and land masses, marine plants receive about 90 x 10^{21} cal/year and terrestrial green vegetation receives about 25 x 10^{21} cal/year. If we assume photosynthetic efficiency is 2% of this energy, it amounts to 50 x 10^{9} tons of organic carbon for terrestrial green plants and 180 x 10^{9} tons for marine plants.

Rabinowitch (1951), calculates that photosynthesis produces approximately 155 x 10^{9} tons of organic carbon of which 90% is contributed by the oceans. However, some of the recent estimates (Rhyer and Wood Well, 1970) suggest a ratio of 1:3 fixation of carbon between terrestrial and marine plants. Of the total amount of light received on earth's surface (650 x 10^{21} cal/year), about 60% is in the form of very short and very long wavelength (far and infra red) radiation and does not participate in photosynthesis. This leaves a total of 180 x 10^{21}cal/year for oceans and 80 x 10^{21} cal/year for terrestrial area. Added to this, is the fact that more than half of the land area is devoid of vegetation. As a result (efficiency of photosynthesis being 2%), this represents only 0.5 x 10^{21} cal/years. Sheoder estimated the production of organic carbon by land plants to be about 16 x 10^9 tons, while Leibeg and Ebermayer estimate the carbon production by land plants at 30 x 10^9 tons and 24 x 10^9 tons respectively. In oceans, where planktons and other plants absorb about 50% of the incident light, the estimates come to 1.8 x 10^{21} cal/year or 180 x 10^9 tons of organic carbon.

HISTORICAL ACCOUNT

Eventhough scientific study of photosynthesis is very recent, from time immemorial man has looked upon the Sun as the source of ale light and sustenance. Early writings of Indians, Romans and Greeks bring out the idea very clearly. The Rig veda says that the 'Sun is the soul of the entire world'. *Bhagavadgita* also mentions that all forms of life are dependent on the Sun. Apart from these early writings, the following may be mentioned as significant names in the development of photosynthetic knowledge.

Aristotle (257 - 80 B.C.), **Theoprastus** (370 - 285 B.C.), and others up to the 17th century believed that plants obtained their nutrition from the soil, and the Sun had nothing to do with plant nutrition.

Van Helmont (1648)

In this experiments with willow seedlings, he observed that water, and not the soil was the principle source of nutrition.

Wood Ward (1699)

Opined that the plant body is made up not of water, but a peculiar soil substance which is absorbed along with water.

Stephen Hales (1927)

Who is regardes as the **Father of Plant Physiology** proposed that green plants obtained their nutrition partly from the air through the leaves, and sunlight may have a role in it. He, however, was not aware of the part played by carbon dioxide.

Joseph Priestley (1972)

Who conducted several experiments on mint plants stated that plants can purify air. He also observed that impure air (CO_2) would get purified (vital air - O_2) if kept in contact with the green mint plants. Priestley also knew nothing about the role of either CO_2 or light in photosynthesis.

Jan Ingen Housz (1771)
A Dutch physician, demonstrated that plants purified air only in the presence of light. He also stated that plants too contributed to 'bad air' during darkness.

Jean Zenebier (1782)
A Swiss minister agreed with the findings of Ingen Housz, and commented that plants absorb CO_2 from the atmosphere during daytime and exhale it (CO_2) during darkness. The role of two gases (O_2 and CO_2) had been implicated in plant nutrition by 1782. Works of Lavoisier and others identified that these gases were O_2 and CO_2.

N.T. de Saussure (1804)
Identified that water is an active participant in photosynthesis together with CO_2. He also made the first quantitative measurements of photosynthesis and opined that approximately equal volumes of CO_2 and O_2 are exchanged during photosynthesis.

Mayer (1842)
Who formulated the law of conservation of energy, observed that during photosynthesis solar energy is stored in the form of chemical energy.

In 1871, two French Chemists, Joseph Bienaime Caventou, and Piere Pelletier isolated the chlorophyll pigment from green plants. Richard Willstatler obtained chlorophyll in the pure form. He also identified two forms – **Chlorophyll a** and **Chlorophyll b** which differed in the manner of their light absorption.

Leibeg (1843)
Stated that the source of all carbon in plants is the CO_2 of the atmosphere.

Julius Sachs (1846)
A German plant physiologist, observed the growth of starch grains in choloroplasts, only in the area of the leaf exposed to light. Thus, he was the first one to note the increase in organic content due to photosynthesis.

Boussingault (1846)
Calculated the ratio of CO_2 intake and O_2 release, and came to the conclusion that it is 1:1.

Baeyer (1870)
Proposed the formaldehyde hypothesis. He argued that CO_2 gets reduced and

forms CH_2O which on polymerization forms the glucose molecule. The formaldehyde hypothesis, however, has been abandoned not only because it has not been detected in plants, but it is also poisonous (to the plants), if it is present.

Blackmann (1905)
Identified that photosynthesis involves two distinct phases if it is present.

Blackmann (1905)
Identified that photosynthesis involves two distinct phases *viz.*, light phase and dark phase.

Van Niel (1931)
Showed that photosynthesis is a biological oxidation reduction reaction.

Robin Hill (1937)
Showed that photosynthesis is localized in the chloroplasts. In the presence of certain electron acceptors (oxidizing - reducing agents), isolated chloroplasts accept light energy and release O_2. This phenomenon of illuminated (isolated) chloroplasts splitting water and releasing O_2, has come to be known as Hill's reaction.

Samuel Ruben, Randall Kamen and Hyde (1941)
Showed (using radioisotopes of oxygen) that the oxygen released during photosynthesis comes only from water and not from CO_2.

Consden, Gordon and Martin (1944)
Developed paper chromatography technique for the separation of amino acids. This technique later was employed to detect the various intermediate compounds during the dark fixation of CO_2.

Arnon (1951 - 60)
Developed the concept of photophosphorylation and showed that coenzyme II picks up the H released by splitting up of water.

Benson, Bassham, Melvin Calvin and others have worked out in detail the carbon reduction cycle with the help of techniques like chromatography and radioautography.

RAW MATERIALS FOR PHOTOSYNTHESIS

There are some basic requirements (raw materials) for the process of photosynthesis. These are CO_2, a hydrogen donor (water), and light energy, besides of course the structural framework of the green plant in the form of chloroplasts.

Carbondioxide

The only source of carbon for plants for the purpose of assimilation is the atmospheric carbon dioxide. Approximately about 20 x 20^{11} tons of carbon dioxide is used for the production of 50 x 10^9 tons of organic carbon. Besides atmosphere, other sources of carbon are biological respiration decaying organic matter etc. Carbon dioxide is in the gaseous form in the atmosphere, while it is in dissolved condition in water.

Carbon dioxide is present in trace quantities in the atmosphere (o.0-35%). There has been a gradual increase in the concentration of the atmospheric carbon dioxide since the middle of 19th century perhaps due to increased pollution (Wood Well, 1970).

Hydrogen Donor

In almost all photosynthetic organisms, water is the hydrogen donor. Water splits into H and OH ions in the presence of light. H is picked up by the enzyme nadph for reduction of CO_2, while OH recombines to form water and oxygen.

Certain chemosynthetic bacteria, however, use compounds of Iron and Sulphur as hydrogen donors.

Light

Solar radiation is the only source of light for all forms of life. The radiation from Sun is emitted in the form of electromagnetic radiation with radiowaves at one end and cosmic rays at the other. Much of this radiation light shielded by the outer reaches of the atmosphere and only visible Light may be defined as the visible (to human eye) part of the electromagnetic spectrum. The visible spectrum is confined between wavelengths of 390mμ and 760 mμ. Spectral analysis of the visible light shows the following rays with their respective wavelengths given in parenthesis.

Violet	(390 - 430 mμ)
Blue	(430 - 470 mμ)
Bluegreen	(470 - 500 mμ)
Green	(500 - 560 mμ)
Yellow	(560 - 600 mμ)
Orange	(600 - 650 mμ)
Red	(650 - 760 mμ)

Rays of the radiation with wavelengths shorter than violet are called U.V. (Ultra violet - 100 A^0 - 390 A^0), X rays (0.1 A^0 - 100 A^0), Gamma rays (0.001 A^0 - 0.1 A^0) and cosmic rays (less than 0.001 A^0). Radiation with wave lengths longer than red are called infra red (7600 A^0 - 1000,000 A^0). Beyond this, are electric and radio waves, which are measured (in terms of wavelength) in kilometers.

Radiation travels in the form of discrete corpuscles or particles called **Quanta** or **Photons.** Each quantum represents a unit of energy. Energy quantity of each ray depends on its wavelength; shorter the wavelength, higher is the energy. The wavelengths of radiation are so small that they are measured not in terms of millimeters or centimeters, but in terms of microns (μ), millimicrons (mμ) or nanometers (nm) or Angstrom (A^0) units.

$1\ A^0 = 10^{-10}$ or 10^{-8} cms.

Each photon has an energy constant $E = hc/\lambda$, where h is planck's constant (=6.625 x 10^{-27} erg/sec) c is velocity of light (3 x 10^{10} cms/sec), and λ is the wavelength as given by Einstein.

The energy value of different light rays is given in the following table.

Energy Value of Different Wavelengths of Light

Sl.No.	Wavelength in mm	Approximately colour	Energy in cal
1.	300	Ultra Violet	95,200
2.	420	Violet	68,000
3.	470	Blue	60,760
4.	530	Green	53,890
5	620	Orange	46,060
6.	700	Red	40,800
7.	683	Red	41,810
8.	672	Red	42,490
9.	650	Red	43,940
10.	850	Infra Red	33,590
11.	1000	Infra Red	28,560

THE PHOTOSYNTHETIC APPARATUS

Within the cell of the green plants are present certain unique organelles called plastids. These perform many vital functions in cell physiology. Among the plastids, the most important from the point of view of photosynthesis are chloroplasts. All the physiological reactions, starting from the absorption of light energy to the fixation of CO_2 take place in the chloroplasts. No wonder then, chloroplast is aptly called the photosynthetic apparatus.

CHLOROPLAST

These are by far, the commonest and the most important among plastids. As the primary sites for trapping and converting solar energy, they are very vital for the existence of not only the green plants, but for the whole living world.

Shape

Chloroplasts have varied shapes. The algal chloroplasts exhibit an array of shapes. They may be cup shaped (*Chlamydomonas*), girdle shaped

Size

There is a great variety in size. Normally, they are about 1 μm thick and 4–6μm in length. Chloroplasts of polyploid cells are generally larger than in the diploid cells. Cells in the shaded area have longer chloroplasts but with less intensity in colouration.

Distribution

They are uniformly, distributed all over cytoplasm, but in some instances they cluster towards the nucleus. The concentration of chloroplast will also depend on light intensity.

Number

The number of chloroplast per cell varies. In some algae, like *Chlamydomonas,* there is only one chloroplast per cell, or two as in

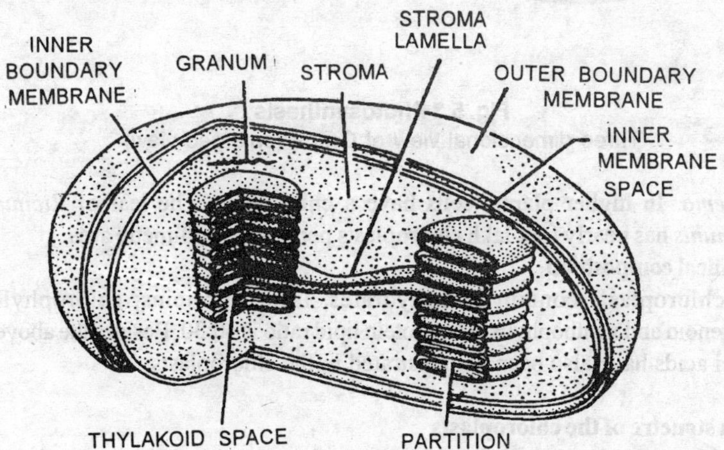

Fig. 5.2 Photosynthesis
Model of Ultrastructure of chloroplasts

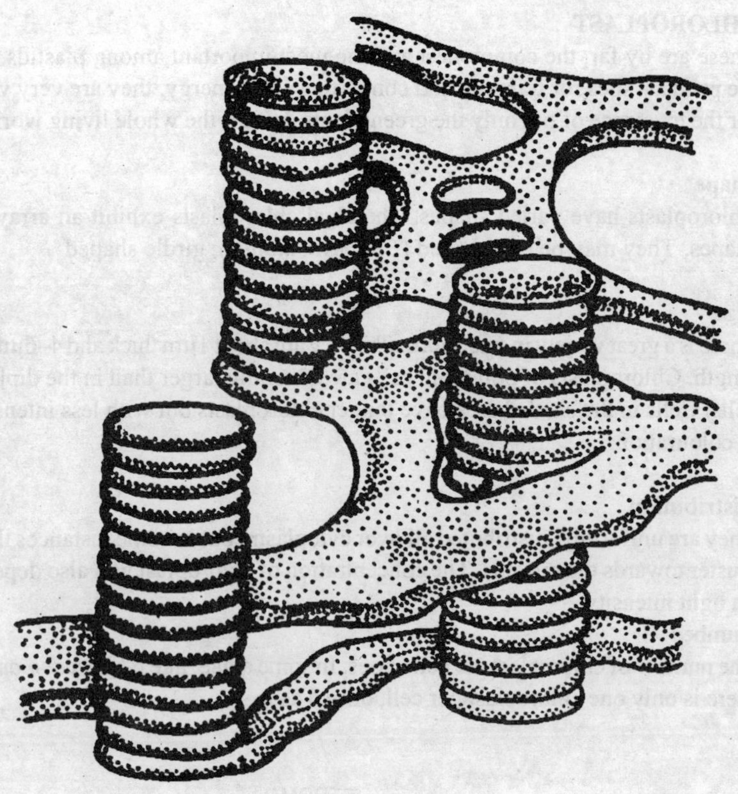

Fig. 5.3 Photosynthesis
Three dimensional view of Grana (enlarged view)

Zygnema. In higher plants, cells have a chloroplast. The leaf of *Ricinus communis* has nearly four lakh chloroplasts per square millimetre area.

Chemical composition

The chloroplasts contain carbohydrates, lipids, proteins, chlorophyll, Carotenoid and xanthophylls. The protein lipid ratio is 40:30. Besides the above, nuclei acids have also been demonstrated in the chloroplasts.

Ultra structre of the chloroplasts

The chloroplast has a covering of two membranes with an inner membrane space. These membranes are smooth and there are no perforations or particles. The membranes are differentially permeable (Mudrak 1969).

A section of the chloroplast reveals an intricate system of membranes enclosed in a granular matrix. These membranes are called *lamellae* and the surrounding matrix – the *stroma.* In a sectional view, the lamellae can be seen packed and these stacks are called *thylakoids.* In the chloroplasts of higher plants, the thylakoids themselves form highly compact bundles called **grans.** Some thylakoids of granum extend into the stroma and maintain contact with other grana. These are called – Stroma thylakoids or stroma lamellae or intergrana.

The chloroplasts of algae lack the granum arrangement and in cyanobacteria (blue green algae), the thylakoids lie naked in cytoplasm without any envelope. In such instances, pigments are uniformly distributed on or in the lamellae.

The stroma has in addition to the lamellae – granules (globuli), lipid droplets, starch grains and vesicles.

Structure of the lamellae

All the photosynthetic pigments are concentrated in the chloroplast, lamellae as well as the outer membranes are composed of lipoprotein, sub units. Each sub unit according to Weir and Benson, has a protein core surrounded by a lipid sheath. The chloroplast membranes have a single layer of sub units, whereas in thylakoids, they are double due to the appressing of two membranes. The appressed areas are called partitions. These are hydrophobic in nature. The space between the two membranes of a thylakoid (fret membrane), enclose a space called fret channel. The space between two thylakoids is called loculus while the end of disc shaped thylakoids are called margins. These areas (loculus, fret channel) are hydrophilic. The chlorophyll molecules are concentrated in the fret cannels i.e., in the space between the two membranes of thylakoids. The chlorophyll molecules may be entirely in space, or the heads may be in the fret channel, while the tails are buried into the sub units. Occassionally, the entire chlorophyll molecule may be found in the fret membrane. Other pigments like cytochromes, carotenoids etc are also found in the fret membranes.

Four sub units of the partition consitute a photosynthetic unit or a *quantosome.*

Pigments in chloroplast

Some of the important pigments present in chloroplasts are chlorophylls, Carotenoids, cytochromes etc. In algal cells, the thylakoids also have phycobilins.

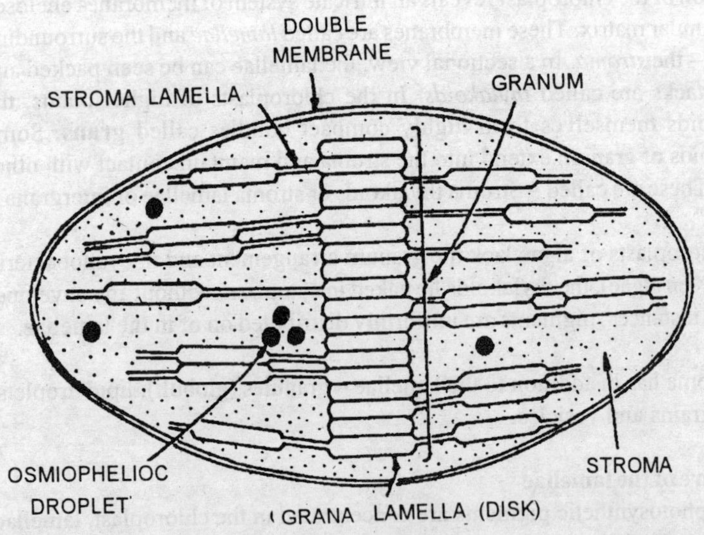

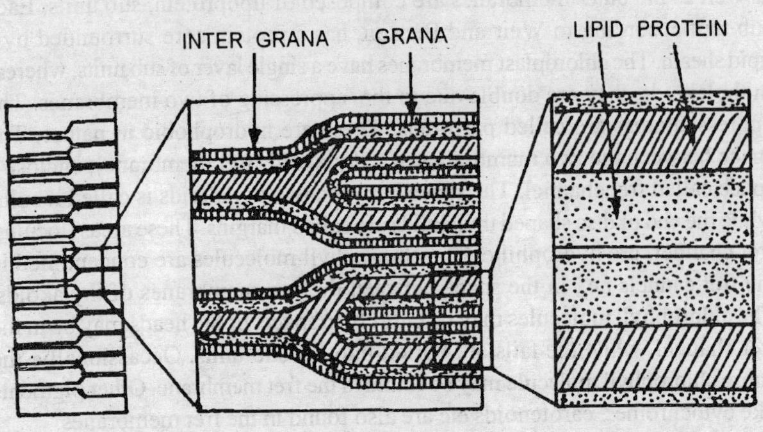

Fig. 5.4 Photosynthesis

Diagrammatic representation of Ultrastructure of chloroplast showing the lamella, **A.** Entire chloroplast, **B-D.** Show successive enlargement of a part of the granum

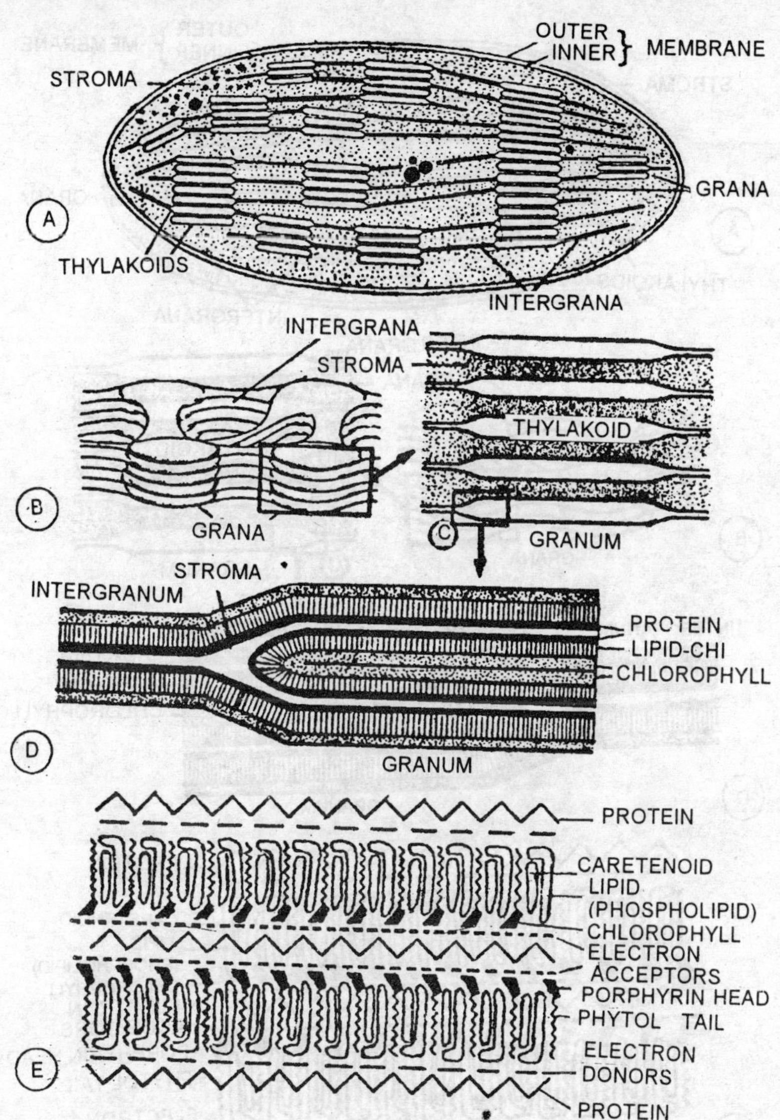

Fig. 5.5 Photosynthesis

Stages in the development of chloroplast**A.** Entire chloroplast, **B.** Grana and Intergrana (diagrammatic), **C.** Section through a granum, **D.** Lamellar structure (in detail), **E.** Model for the arrangement of pigments in a granum

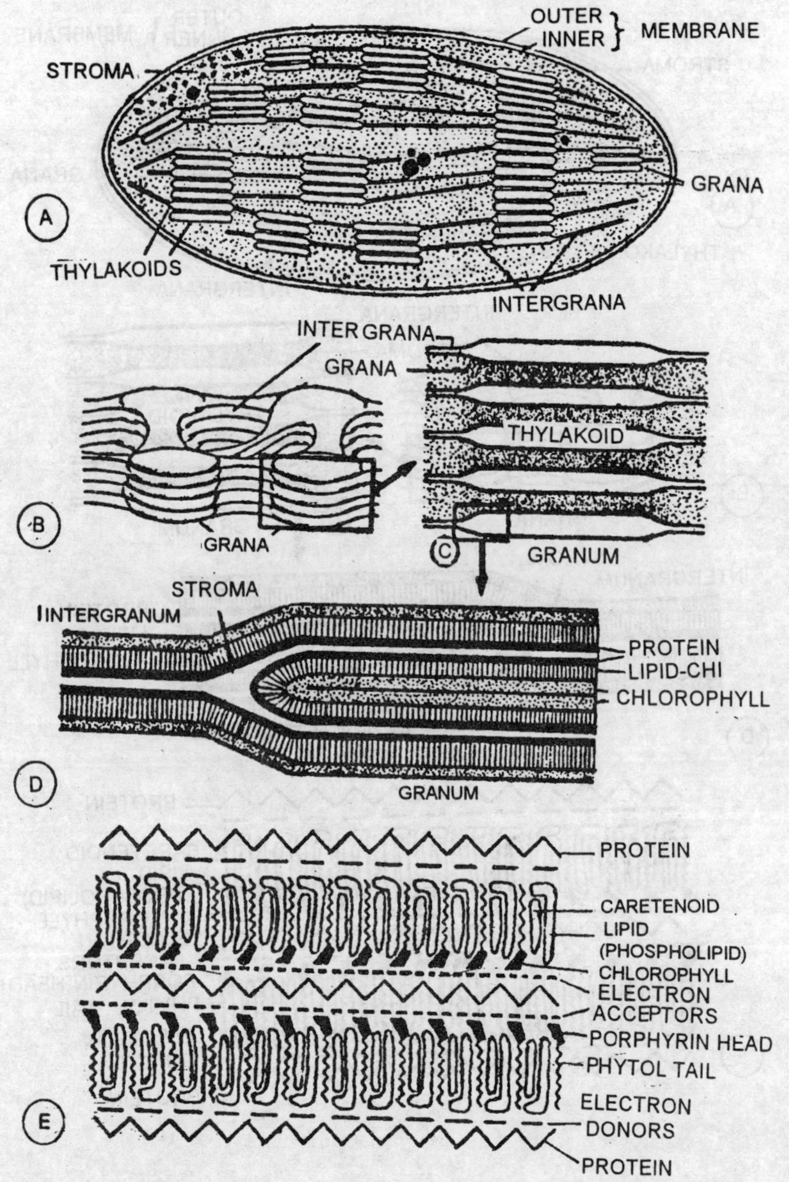

Fig. 5.6 Photosynthesis

A. Entire chloroplast, **B.** Grana and Intergrana (diagrammatic), **C.** Section through a granum, **D.** Lamellar structure (in detail), **E.** Model for the arrangement of pigments in a granum

Chlorophylls

These are the most important pigments involved in the basic photochemical reaction that sustains life. Atleast nine types of chlorophylls have been distinguished so far – chlorophylls *a,b,c,d* and *e.,* bacterio chlorophylls *a* and *b,* and the chlrobium chlorophylls 650 and 660 µm.

Among the chlorophylls, chlorophylls, and b are best known and most widely distributed. These are absent in pigmented bacteria. Chlrophylls *c*, *d* and *e* are found only in algae in combination with chlorophyll a. Bacterio chlorophylls and chlorobium chlorophylls are found only among photosynthetic bacteria.

The chlorophyll molecule is a complex structure basically made up of a 'head' and a 'tail' resembling a tennis racquet. The head is a porphyrin structure made up of four Pyrrole rings attached to each other at the centre by an isocyclic ring containing Mg atom at the centre. Extending from one of the pyrrole rings is the tail–the alcoholic chain (Phytol). The emperical formula for the chlorophyll molecule is $C_{55}H_{72}O_5N_4$ Mg. The phytol chain is esterified on the C atom of one of the pyrrole rings and has only one double bond.

Chlorophyll *a* and *b* differ structurally in having different atomic groupings at the C_3 atom in the pyrrole ring. In chlorophyll *a,* C_3 atom has a methyl group, while chlorophyll b has an alehyde group; besides the two pigments have different absorption spectra. The peaks are as follows.

Chlorophyll *a* 429, 410, and 66nm

Chlorophyll *b* 430, 453 and 442 nm

(The above spectra are of chlorophyll *in vitro*)

Carotenoid pigments

These are lipid compounds ranging in colour from yellow to purple. They are widely distributed in both plants and animals. They are also present in microorganisms including red algae, cyanobacteria, photosynthetic bacteria, fungi etc. Wackenroder (1931), isolated the first Carotenoid – carotene from, the carot tissue.

Carotenoids are derivatives of lycopene, a red pigment found in many plants. They (Carotenoids) are known to have eight isoprent (CH2=C [CH3] - CH = CH2) like residues.

The main Carotenoid found in plants is the orange yellow coloured β carotene with some quantity of α carotene.

The Carotenoids are also located in the chloroplast. According to Goodwin

(1960), Carotenoids and chlorophylls may be combined with the same protein to form a complex known as photosynthein.

Carotenoids protect the chlorophyll from photooxidation and transfer the light energy they absorb to chlorophyll *a*.

Phycobilins

These are found only in algae. There are two known types of phycobilins – the phycoeroythrin (red), and phycocyanin (blue). These pigments are strongly associated with a protein and consequently it is difficult to isolated the pigment in the pure state.

Like Carotenoids, phycobilins are also involved in the transfer of light energy (they absorb) to the chlorophyll.

The absorption spectra of phycobilins presents an interesting study considering the fact that they act as accessory pigments in photosynthesis. R - phycoerythirn has peaks at 495, 540 and 545 nm, while R-phycocyanin has peaks at 550 and 615 nm.

Photosynthetic units

A photosynthetic unit may be defined as "the smallest group of coordinating pigment molecules necessary to effect a photochemical act, i.e. absorption and transportation of a light quantum to a trapping center where it causes excilation and release of an electron".

Emerson and Arnold (1932), who experimented in Chlorella plants, observed that about 2,500 molecules of the chlorophyll pigment are necessary to fix a molecule of CO_2 and they termed this number (2,500), as a photosynthetic unit. They also observed that to release one molecule of O_2, 10 quanta of light are necessary. Some of the recent works, however have shown that on the basis of absorption of 10 quanta, photosynthetic unit should consist of 230 molecules of the chlorophyll pigment. Park and Beggins (1964), working on the chloroplasts opined that photosynthetic units are distinct morphological identities and called them quantasomes. The quantasomes are 180 A^0 broad and 100 A^0 thick (`8 x 16 x 10nm). Each quantasome has 230 molecules of chlorophyll (160 chlorophyll a and 70 chlorophyll b), 48 carotenoids, 46 quinone compounds, 116 phospholipids, 144 digalactosyl diglycerides, 346 monogalactosyldiglycerides, 48 sulfolipids, unidentified lipids and many sterols. Quantasomes have a molecular weight of about two millions and are arranged in a monolayer on the thylakoid bodies. Electron microscopic studies have revealed that the individual quantasome may have four or more sub units. This middle region of the quantasome is called the **reaction centre.**

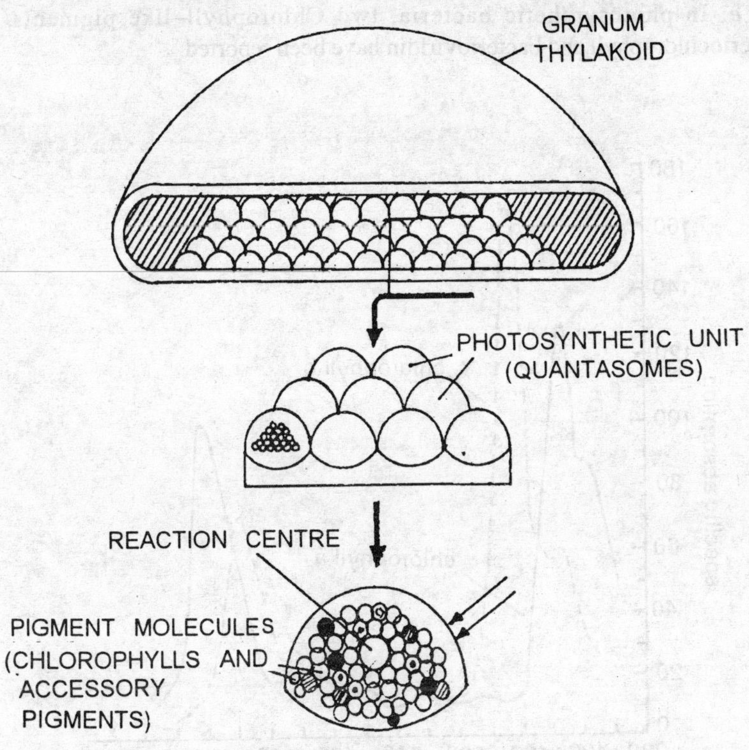

Fig. 5.7 Photosynthesis
Diagrammatic representation of a thylakoid, quantasomes and the reaction centre

Robert Huber, Johann Deisenhofer, and Jartmutt Michel were awarded the 1988 chemistry Nobel Prize for unravelling the secret of the reaction centre. Working at the Max Planck Institute in Martinsried, Germany, these scientists, made their breakthrough by extracting the reaction centre from the membranes of certain photosynthetic bacteria. They then crystallized it and analysed the structure using X - ray diffraction.

Photosynthetic pigments

The most important pigments involved in photosynthesis are chlorophylls and carotenoids. Other pigments present, but not involved in photosynthesis are biliproteins or chromatoproteins.

Chlorophylls

There are five important types of chlorophylls. These are Chlorophyll *a,b,c d* and *e*. In photosynthetic bacteria, two Chlorophyll–like pigments–bacteriochlorophyll and bacterioviridin have been reported.

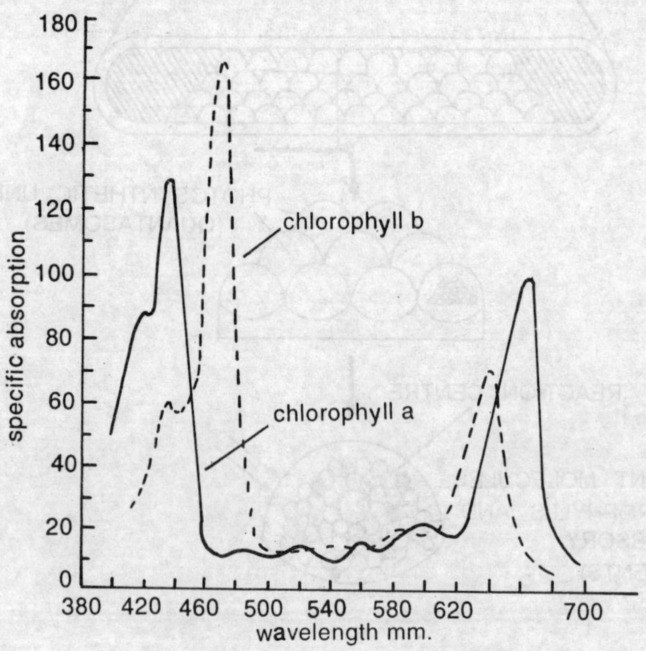

Fig. 5.8 Photosynthesis
Absorption spectra of chlorophyll a and b

Chlorophyll *a* is present both in higher plants and algae. The pigment is insoluble in water but readily soluble in either. It appears blue green in transmitted light, but reddish in reflected light. Chlorophyll *a* is the principal pigment involved in the trapping of light energy.

Chlorophyll *b* is also found in green and higher plants. It is yellowish green in transmitted light but reddish in reflected light. It is similar to chlorophyll a in solubility. Chlorophyll *b* is absent in Cyanophyceae, Rhodophyceae Phaeophyceae and diatoms.

Chlorophyll *c*, *d* and *e* are present in brown, red and golden brown (Xanthophyceae) algae respectively.

Carotenes and Xanthophylls. They are insoluble in water but soluble in organic solvents.

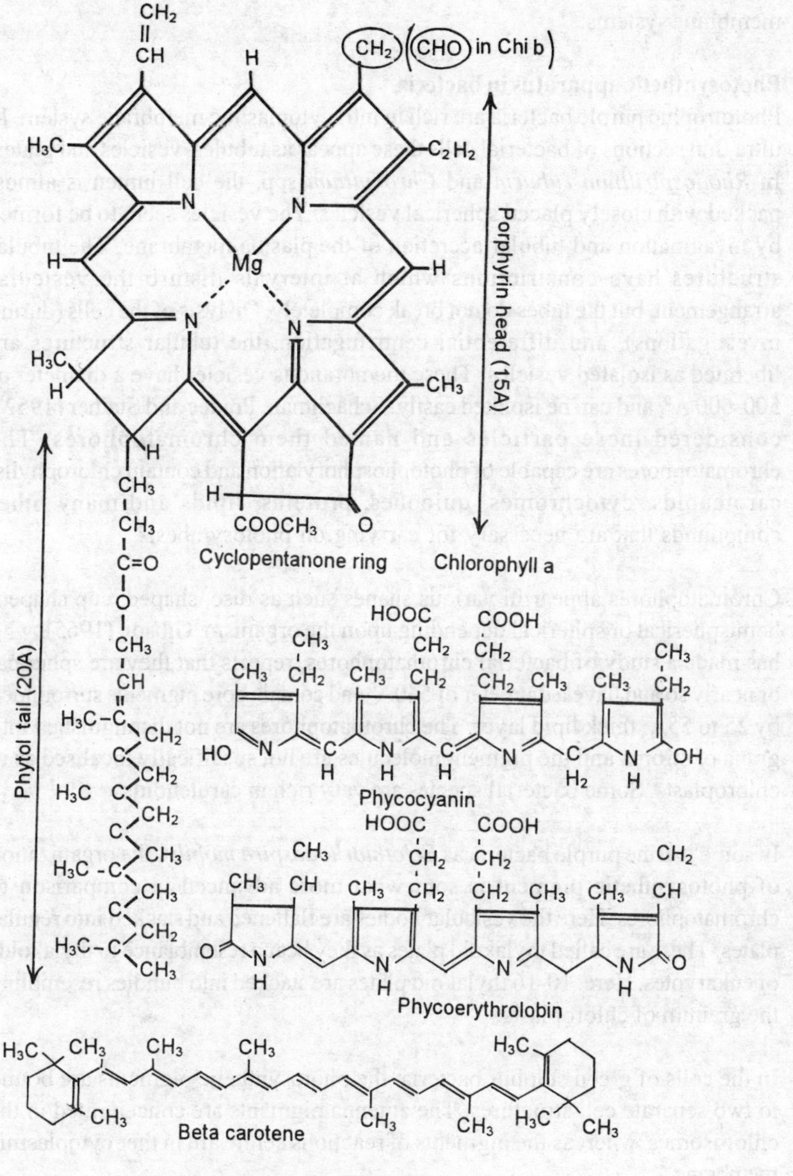

CH_2 (CHO in Chl b)

Porphyrin head (15A)

Cyclopentanone ring

Chlorophyll a

Phytol tail (20A)

Phycocyanin

Phycoerythrolobin

Beta carotene

Fig. 5.9 Photosynthesis
Molecular structure of photosynthetic pigments

Photosynthetic apparatus in prokaryotes

In photosynthetic prokaryotes, the pigments are not localized into definite organelles as in eurkaryotes. But they do have some organization in some membrane systems.

Photosynthetic apparatus in bacteria

Phototrophic purple bacteria are rich in intracytoplasmic membrane system. In ultra thin sections of bacterial cells these appear as tubules, vesicles and plates. In *Rhodospirillium ruburm* and *Chromatium* spp, the cell lumen is almost packed with closely placed spherical vesicles. The vesicles seem to be formed by invagination and tubular accretion of the plasma membrane. The tubular structures have constrictions which at intervals disturb the vesicular arrangement, but the tubes do not break completely. On lysis of the cells (during investigations), and differential centrifugation, the tubular structures are liberated as isolated vesicles. These membranous vesicles have a diameter of 500-600 A^0, and can be isolated easily. Schachman, Pradec and Stainer (1952), considered these particles and named them chromatophores. The chromatophores are capable of photophosphorylation and contain chlorophylls, carotenoids, cytochromes, quinones, proteins, lipids and many other compounds that are necessary for carrying on photosynthesis.

Chromatophores appear in various shapes such as disc, shaped, cup shaped, hemispherical or spherical depending upon the organism. Gibson (1965), who has made a study of bacterial chromatophores, reports that they are spherical or nearly so and have a diameter of $540A^0$ and contain core pigments surrounded by 25 to $55A^0$ thick lipid layer. The chromatophores are not distinguished into grana or stroma and the pigment molecules are not specifically localised as in chloroplasts. Some bacterial species are very rich in carotenoids.

In some marine purple bacteria as *Eclothiorhodospira mobilis,* the organization of photosynthetic pigment is somewhat more advanced in comparison to chromatophores. Here the vesicular bodies are flattened and stacked into regular plates. These are called thylakoid plates as they bear a resemblance to thylakoids of eukaryotes. Here, 10-16 thylakoid plates are stacked into bundles resembling the granum of chloroplasts.

In the cells of green sulphur bacteria. the photosynthetic pigments are bound to two separate cell structures. The antenna pigments are concentrated in the chlorosomes, whereas the pigments of reactions centre are in ther cytoplasmic membrane.

The photosynthetic pigments of phototrophic purple bacteria are bound to intracytoplasmic membranes.

Photosynthetic apparatus in cyanobacteria

In cyanobacteria, the peripheral region of cytoplasm contains the photosynthetic apparatus. In unicellular forms, this commonly consists of a few membrane layers which extend in concentric sheets below the cell membrane and surround the central nucleoplasmic zone. In filamentous members, the cells have more elaborate membrane systems formed due to expansion and invagination of the membranes. The basic structure of the membranes looks like a flattened sac and is designated as the thylakoid. The membranes which are closely thylakoid sac.

The thylakoid lamellae have a lipid bilayer incorporating the lipophilic photosynthetic pigments – chlorophyll a and various carotenoids. In addition, it also incorporates cytochromes, plastocyanin and other components of the electron transport chain. Chlorophyll *a* which is the only chlorophyll present in cyanobacteria exists in three spectral variations (with peaks at 670, 680 and 700nm). Among the carotenoids, which absorb in the range between 480 and 520nm, β carotene appears to be widely distributed in cyanobacteria. Xanothophylls, however, differ in their abundance as well as presence in various species of cyanobacteria. Some of the xanthophylls seen in cyanobacteria are cchincone, zeaxanthin, oscillaxanthin, myxoxanthophyll etc.

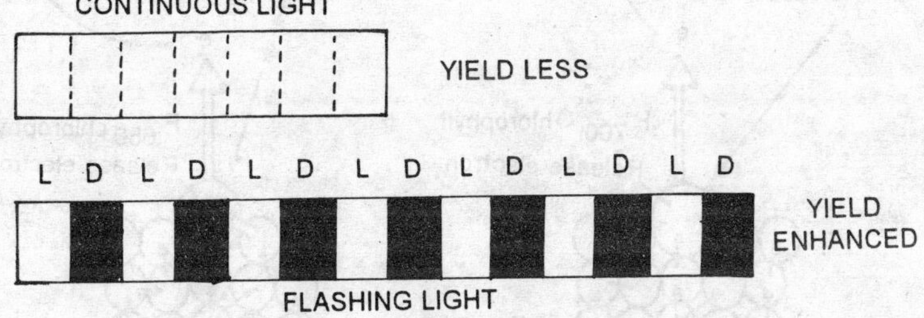

Fig. 5.10 Photosynthesis

Red drop and Emerson effect in Chlorella

A significant part of the photosynthetic pigment complement of cyanobacteria is located in some granular supermolecular complexes such as phycobilisomes. These contain the chromophore bearing water soluble phycobiliproteins namely blue phycocyanin, allophycocyanin and phyceorythrins. The relative quantities of these pigments may vary according to species and the spectral composition of light. It is that which determines the varied colour appearance of the organisms.

The table belwo gives an account of the various classes of pigments present in cyanobacteria.

The major photosynthetic pigments of blue - greens.

Group	Class	Major absorption bands (nm)
Chlorophylls	Chlorophylla	435, (670), 680, (700)
Carotenoids	β -carotene	(431), 450-454, 478-480
	Echinenone	455-459, (475)
	Zeaxanthin	(430), 453, 479
	Canthaxanthin	466
	Myxoxanthophyll	504-508, 474-476, 450-452
Phycobiliproteins	Allophococyanin	650
	Phycocyanin	610-625
	Phycoerythrin	555-565

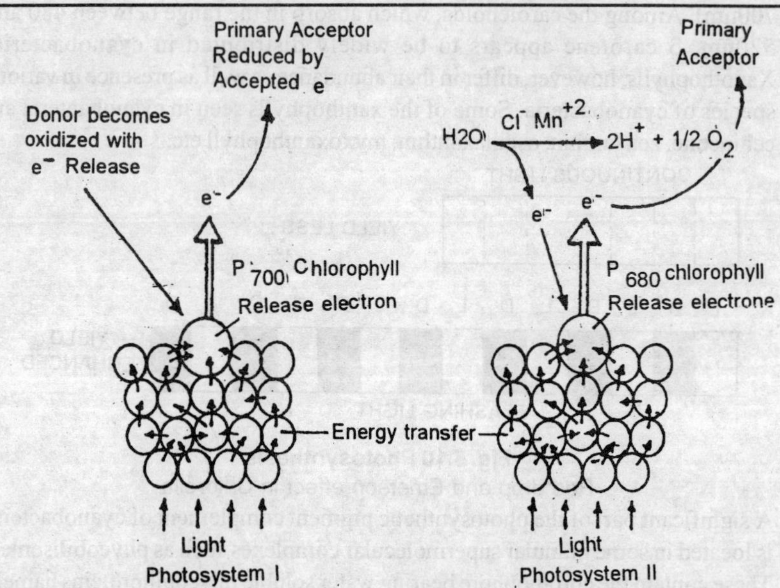

Fig. 5.11 Photosynthesis

Energy transfer in Photosystem I and Photosystem II in green plants

Phycobiliproteins represent, the primary light harvesting or antenna pigments, for photosystem II in cyanobacteria. Light energy trapped by these pigments is transferred to chlorophyll with a very efficiency. Each phycobilisome consists of a triangular core surrounded by six peripheral rods and is made up of disc shaped subunits. The triangular core which is in direct contact with the thylakoid membrane has the pigments allophycocyanin, while the rods contain phycocyanin and phycoerythrin. The phycobilisomes are arranged in parallel rows on the outer surface of the thylakoid membrane. Close proximity between phycobilisomes and thylakoid membranes is necessary for maximum energy transfer in the primary act of photosynthetic.

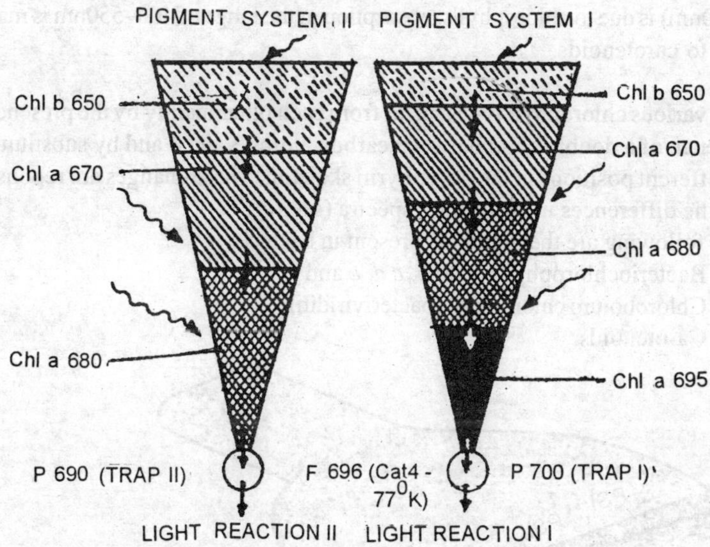

Fig.5.12 Photosynthesis
Distribution of different pigments in PS I and PS II

Complementary chromatic adaption

Environmental conditions such as mineral nutrients, and intensity and quality of light is known to affect the pigment composition in cyanobacteria. It has been suggested that the changes enable the most efficient utilization of available light energy. The synthesis of phycobilin pigments and even the formation of thylakoids is susceptible to environmental influence. Change in pigmentation is largely due to change in phycocyanin phycoerythrin ratio. Certain strains appear distinctly bluish-green when grown under red light because phycocyanin which absorbs minaly red light is present in large quantities. On the other hand, strains grown under green light will appear reddish due to the preponderance of green light absorbing phycoerythrin in light cells. Thus, the organisms produce a pigment which can absorb best the existing light. The shift between red light and green light effects was found to occur at 590nm wavelength. The phenomenon of this shift is called *complementary chromatic adaptation* or *Gauidukov phenomenon* (named after the discoveror). This phenomenon is seen in only those cyanobacteria which synthesize phycoerythrin.

Photosynthetic pigments is bacteria

Suspensions of photosynthetic bacteria appear green, purple, violet, red, brown or salmon coloured to the eye based on the pigment composition. The pigment components can be readily recognised by their absorption spectra of intact cells. Absorption peak in the blue (450nm), and the red as well as infra red (650-1000nm) is due to chlorophylls; absorption in the range of 400–550nm is mainly due to carotenoids.

The various chlorophyll types differ from each other mainly by the presence or absence of a double bond between carbon atoms 3 and 4 and by substitutions at different positions in the poryphyrin skeleton. These changes are reponsible for the differences in absorption spectra (see fig).

The following are the pigments present in bacteria.

1. Bacteriochlorophyll *a* and *b, c d, e* and *g*
2. Chloroboium chlorophyll (bacteriviridin)
3. Carotenoids

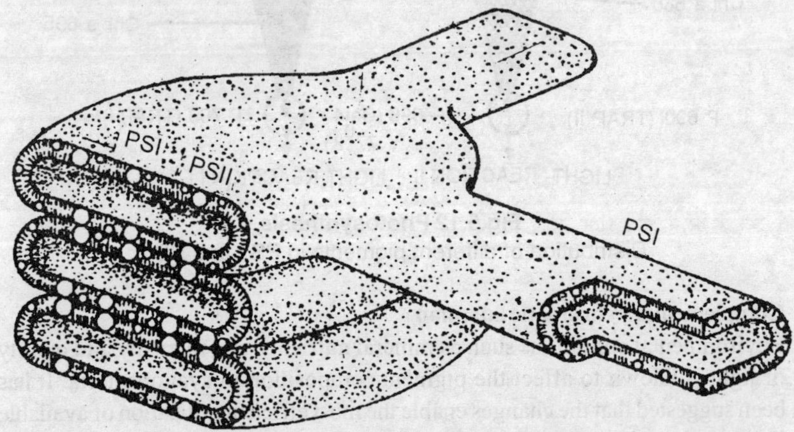

Fig. 5.13 Photosynthesis
Model of the structure of Granum thylakoid and a stroma thylakoid showing the distribution of PS I and PS II

Bacteriochlorophulls

These are found only in purple sulphur and non sulphur bacteria. The types of bacteriochlorophylls differ in their absorption spectra which is given in the following table :

Types of Bacteria Chlorophylls

Designation of Jenson et. Albugo, 1964	Former Designation	Characteristic Adsorption Maxima in living cells (nm)
Bhcl a	Bacteriochlorophyll	375, 590, 805, 830-890
Bhcl b	Bacteriochlorophyll *b*	400, 605, 835-950, 1020-1040
Bhcl c	Chlorobium chl 660	Long wavelength absorption max 745-755
Bhcl d	Chlorobium chl 650	– do – 705-740

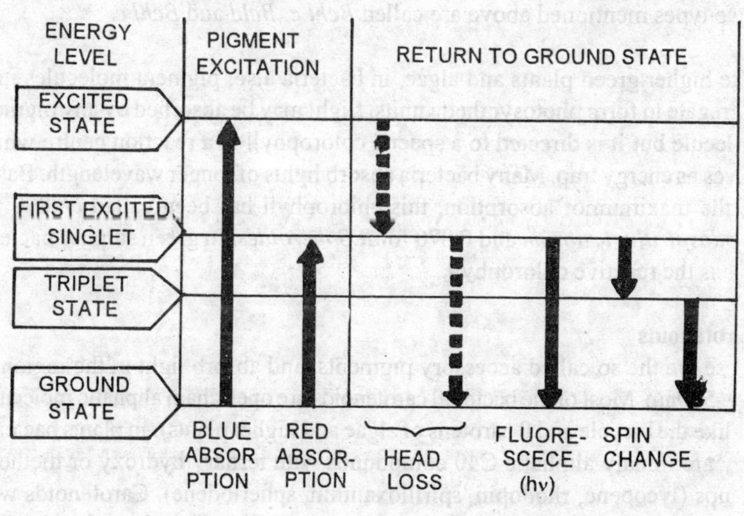

Fig. 5.14 Photosynthesis

Diagram illustrating the absorption of light and changes in the electronic configurations in chlorophyll

Bchl (Bacteriochlorophyll) *a* found in purple sulphur bacteria has a peak between 850–890nm. *Bchl b* which is found in *Rhodopseudomonas viridis*, *Ectothiorhodospira halochloris* and *thiocapsa pfennigii* has an absorption peak at 1020–1040nm.

hl a of purple bacteria occurs in four spectral forms – B 800, B 820, B 850 and 870–890. The differences in absorption spectra are caused by the kind of ding and the position of the *Bchl* molecule in the pigment protein complexes the photosynthetic apparatus. The number of peaks and the height varies tween species and also cultural conditions.

lorobium chlorophyll

ese are present in green sulphur bacteria of the family chlorobiaceae. Three lorophylls– chlorobium chlorophylo 660, chlorobium chlorophyll 650 and orobium chlorophyll 715 have been repored in green sulphur bacteria such *Chlorobium, Chloropseudomonas, Chloroflexus* etc.

cording to the recent nomenclature, however, chlorobium chlorophylls of ee types mentioned above are called *Bchl c, Bchl* and *Bchl e.*

e higher green plants and algae, in bacteria also, pigment molecules may gregate to form photosynthetic units. Light may be absorbed by any pigment lecule but it is directed to a special chlorophyll in a reaction centre which ves as energy trap. Many bacteria absorb lights of longer wavelength. Based the maximumof absorption, this chlorophyll has been called *chl*890 for *odospirillum rubrum* and P 870 for *R.Spheroides.* In green sulphur bacteria *e* is the reactive chlorophyll.

rotenoids

se are the so called accessory pigments, and absorb light in the region of –550nm. Most of the bacterial carotenoids are open chain aliphatic molecules like the bicycle and β carotens of algae and higher plants). In plants bacteria, y are minaly aliphatic C40 compounds with tertiary hydroxy or methoxy ups (lycopene, rhodopin, spirilloxanthin, spheriodene). Carotenoids with matic rings are also found in bacteria. These are restricted to a few families h as chromatiaceae, chlorobiaceae etc.

een bacteria possess monocyclic γ carotenoids. Lycopene is the most widely tributed carotene in bacteria. Bicycle carotenes are rare in photosynthetic teria. The table below gives various groups of carotenoids found in bacteria. rotenoids fulfill two functions. On the one hand, they are photosynthetic on other, they protect the chlorophylls from possible damage due to otoxidative action. It has been seen that carotene free mutants of purple teria can grow only in dim light and are killed by higher light intensities.

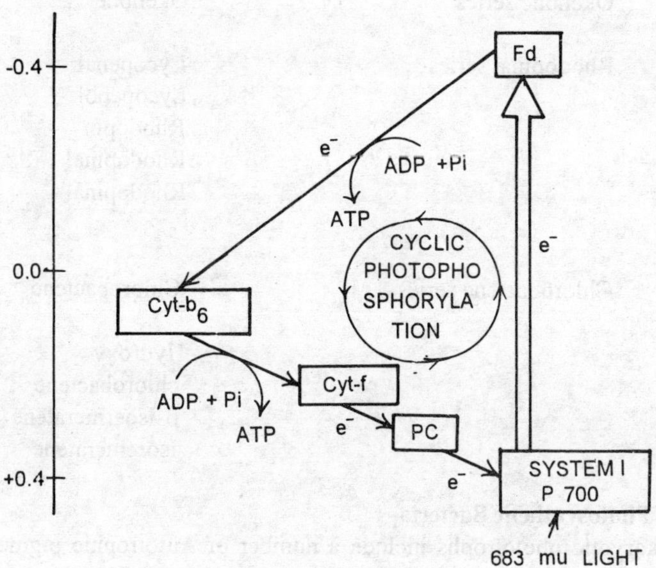

Fig. 5.15 Photosynthesis
Systematic representation of cyclic photophosphorylation

The differences in the absorption spectra of green algae, cyanobacteria, green bacteria and purple bacteria leads to the conclusion that different phototrophic organisms use different spectral rays for their photosynthetic activities. The differences correlate well with the light conditions in their natural habitats. These differences can be used for selective culture of these organisms

Carotenoid groups of phototrophic bacteria

Group	Name	Major components
1.	Normal aspirilloxanthin series	Lycopene
	Rhodopin	
	Spirilloxanthin	
2.	Alternative spirilloxanthin series and ketocarotenoids of spheriodenone type	Spheroidene Hydroxy-spheriodene
	Spheriodenone	
	Hydroxyspheriodenone	
	Spirilloxanthin	

3.	Okenone series	Okenone
4.	Rhodopinal series	Lycopenal Lycopenol Rhodopin Rhodopinal Rhodopinal
5.	Chlorobactene series	Chlorobactene Hydroxy-chlorobactene β-Isoernieratene Isorenieratene

Types of Photosynthetic Bacteria

The prokaryotic phototrophs include a number of Autotrophic pigmented organisms. These are –

1. Purple non sulphur bacteria - Rhodospirillaceae
2. Purple sulphur bacteria - Chromatiaceae
3. Green sulphur bacteria - Chlorobiaceae Chlofoflexaceae
4. Cyanobacteria - formerly blue green algae.

Types of Bacterial Photosynthesis

Phototrophic bacteria normally utilize CO_2 of the environment as a source of cellular carbon. The energy required for the reduction is obtained from inorganic or organic sources. In photolithotrophic organisms, inorganic substances like H_2S or thiosulphate are the source of reducing power. In photoorganotrophic organisms, organic substances like malate or succinate yield reducing power. The general equation for reduction of carbon in photosynthesis is as follows :

$$2H_2A + CO_2 \text{ ——— } C(G_2O) - H_2O + 2A.$$

In algae and other green plants, which grow in the presence of air H_2A is H_2O. In obligatory anaerobic photosynthesizing bacteria. H_2A is H_2S or organic molecules. In chromatiaceae and chlorobiaceae, reductive power comes from H_2S or thiosulphate. In Rhodospirillaceae organic molecules provide reducing power. The following is an outline of different types of reactions involved in bacterial photosynthesis.

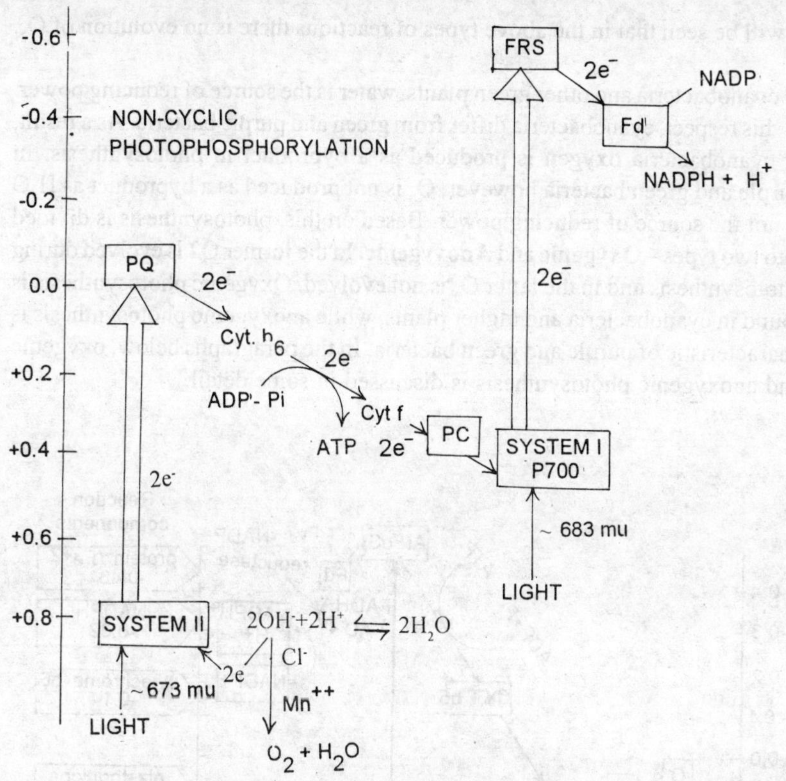

Fig. 5.16 Photosynthesis

Systematic representation of non-cyclic photophosphorylation

1. Purple sulphur bacteria

Here, H_2S is the substrate that provides reducing power. Sulphur is produced instead of oxygen.

$$CO_2 + H_2S \text{ —— } C(H_2O) + 2S + H_2O$$

Purple sulphur bacteria can also utilize thiosulfate as the source of reducing power

$$2CO_2 + Na2S2O_3 + 5H_2O \text{ ——— } 2C(H_2O) + 2NaHSO_4$$

2. Non sulphur bacteria

Here, the reducing power is provided by organic compounds like ethanol, succinate etc.

$$CO_2 + 3C_2H_5OH \text{ ——— } C(H_2O) + CH_3CHO + H_2O$$

It will be seen that in the above types of reactions there is no evolution of O_2.

In cyanobacteria and other green plants, water is the source of reducing power. In this respect, cyanobacteria differ from green and purple bacteria. As a result, in cyanobacteria oxygen is produced as a byproduct in photosynthesis. In purple and green bacteria however, O_2 is not produced as a byproduct as H_2O is not the source of reducing power. Based on this, photosynthesis is divided into two types – **Oxygenic** and **Anoxygenic.** In the former O2 is evolved during photosynthesis and in the latter O_2 is not evolved. Oxygenic photosynthesis is found in cyanobacteria and higher plants, while anoxygenic photosynthesis is characteristic of purple and green bacteria. In the paragraphs below, oxygenic and anoxygenic photosynthesis is discussed in some detail.

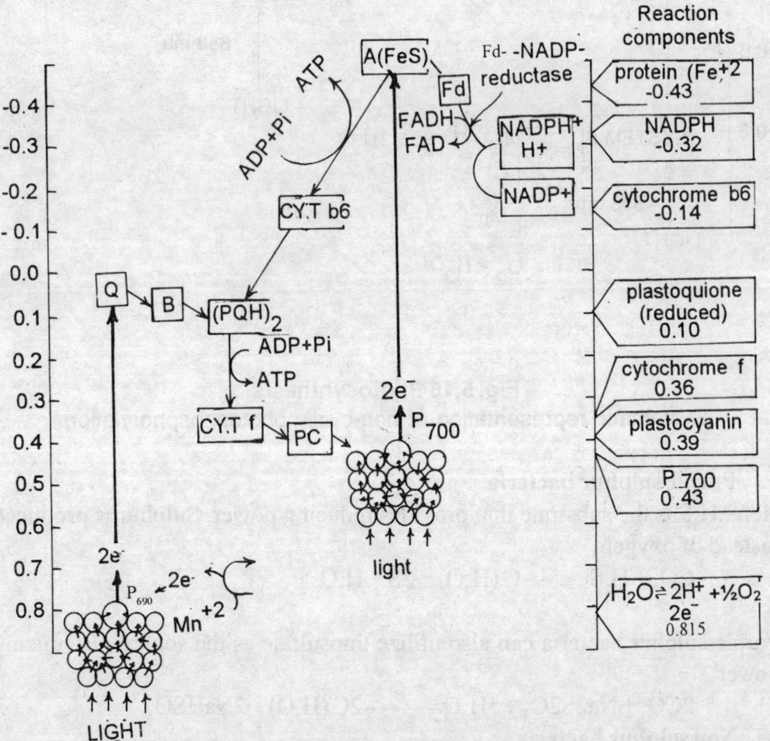

Fig. 5.17 Photosynthesis

Systematic representation of overall electron flow in light reaction combining both cyclic and non-cyclic photophosphorylation

MECHANISM OF PHOTOSYNTHESIS

Photosynthesis in both green plants and bacteria can be understood dividing the process into two–light reaction and dark reaction. In light react molecule(s) of ATP is generated along with reducing power. While AT generated in all phototrophic organisms using light energy as the source reducing power comes from a variety of sources, inorganic compound water. It is based on the source of reducing power (H_2O or otherwise) released or not. Photosynthesis is divided into two types – anoxygenic oxygenic.

Light reaction in anoxygenic photosynthesis

Purple sulphur, and green sulphur bacteria conduct anoxygenic photosynthe Light reaction here involves the following steps.

1. Source of electron
2. The electron transport system
3. Photophosphorylation

Source of electron

Evolution of oxygen has not been demonstrated in bacteria, and hence th are incapable of using water as a source of reducing power. The source electron are therefore, hydrogen sulphide and reduced inorganic and orga compounds.

In sulphur bacteria oxidize sulphide to sulphate and the reducing power utilized to reduce CO_2 as follows :

$$CO_2 + H_2S \longrightarrow 3\,CH_2O + 2S$$
$$3CO_2 + 2S + 5H_2O \longrightarrow 3CH_2O + 2H_2SO_4$$

Chromatium Okenii, a purple bacterium can utilize a variety of sulph compounds and molecular hydrogen.

Electron transport system

In photosynthetic bacteria, the electron transport system consists of a prima electron acceptor (x), a secondary acceptor (y), and *b* and *c* type cytochrom Cytochrome *a* is absent as molecular oxygen does not participate in the electro transport. Some scientists believe that there are two separate electro pathways–one for photosynthesis and another for respiration. But as is see in *R.Rubrum,* only one electron transport mechanism is considered enough f both the processes.

The electron transport carriers are asymmetrically located in the membrane to help maintain the hydrogen ion gradient.

The primary electron acceptor (x) receives the electron from the bacteriochlorophyll P870. Eventhough the exact nature of the primary acceptor is not known, researches have shown that it contains nonhaeme iron and it is quite likely that it is ferredoxin, an iron protein complex.

The events in electron movement can be summarized as follows. The electron given from bacteriochorophyll is accepted by ferredoxin, which then releases an electron to be accepted by FMN (a flavoprotein). From FMN, the electron travels down and is accepted by NAD which gets reduced providing the necessary reducing power (to be utilized for reduction of CO_2). Ubiquinone may receive an electron either from FMN or FAD, and the electron released from it is accepted by *cyt b* and *cyt c*. Finally, the electron moves back to *Bchl*.

Photophosphorylation

The process of conversion of light energy into chemical energy is called photophosphorylation. Two types of phosphorylation are known to exist in bacteria in the same way as in green plants. These are cyclic phosphorylation and non-cyclic phosphorylation. There is one difference between bacteria and green plants. In the case of green plants, the photosynthetic units have two reaction centres to trap light rays of two different wave lengths. In the case of bacteria, however, there is only one reaction centre and only one PS (photo system) is involved.

It was thought earlier that in bacteria there is only cyclic photophosphorylation. This was based on the assumption that evolution of oxygen is a must in non-cyclic photophosphorylation. But some discoveries have clearly shown that non cyclic photophosphorylation occurs in bacteria and that evolution of oxygen need not necessarily accompany non-cyclic photophosphorylation.

Works of Losada, Whatley, and Arnon (1961), have shown that non-cyclic photophosphorylation reaction does not depend on oxygen evolution. The difference between cyclic and non-cyclic photophosphorylation is based on the dependence of the latter on the presence of an electron donor and acceptor which is necessary, produce the reducing power. Occurrence of non-cyclic photophosphorylation in the chromatophores of *Rhodospirillum rubrum* has been shown by suppressing cyclic photophosphorylation with actinomycin *a* and supplying DPIP ascorbate as an electron system, and NAD as electron acceptor.

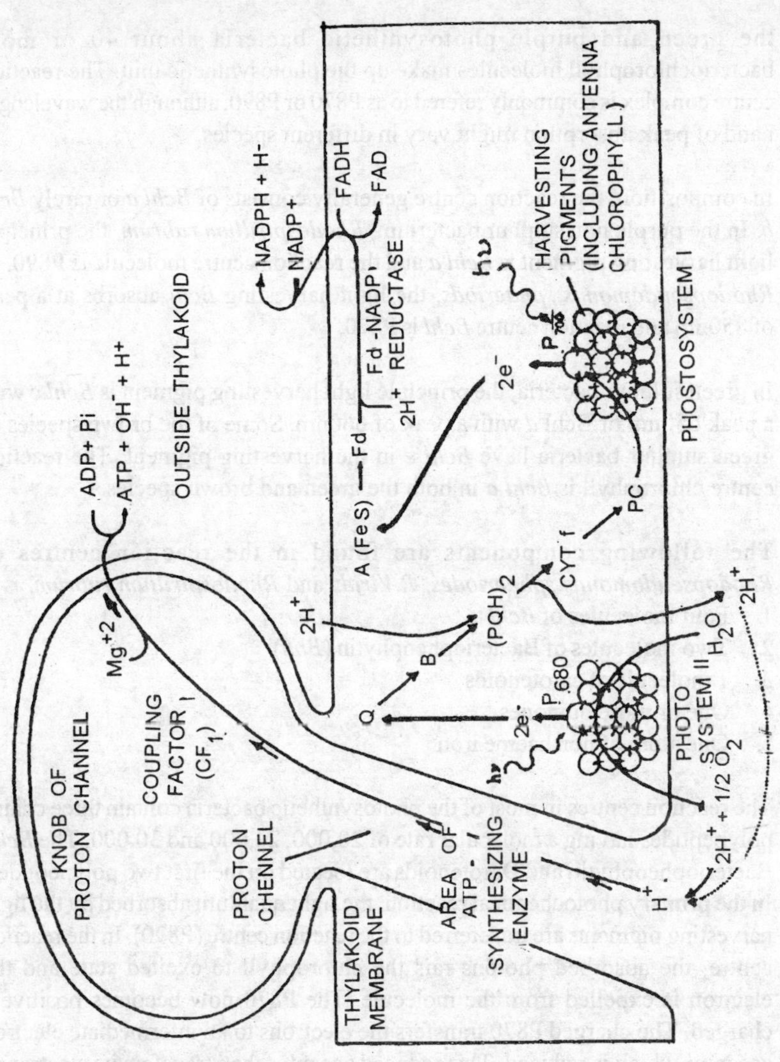

Fig. 5.18 Photosynthesis

Photophosphorylation and ATP synthesis in granum thylakoid according to Chemiosmotic hypothesis

In bacterial pigment system also, the photosynthetic units consists of two parts namely: an *antenna* consisting of light harvesting molecules and the reaction centre where the actual energy conversion takes place. Light energy harvested by the antenna molecules is channelised to the reaction centre. In the green and purple photosynthetic bacteria about 40 or more bacteriochlorophyll molecules make up the photosynthetic unit. The reaction centre complex is commonly refered to as P870 or P890, although the wavelength band of peak absorption might vary in different species.

In composition, the reaction centre generally consists of *Bchl a* or rarely *Bchl b*. In the purple non sulphur bacterium, *Rhodospirillum rubrum,* the principle light harvesting pigment is *Bchl a* and the reaction centre molecule is P890. In *Rhodopseudomonas sphaeriods,* the light harvesting *Bchl* absorbs at a peak of 850nm and reaction centre *Bchl* is P870.

In green sulphur bacteria, the principle light harvesting pigment is *Bchl c* with a peak 650nm or Bchl *d* with a peak of 660nm. Some of the brown species of green sulphur bacteria have *Bchl e* in the harvesting pigment. The reaction centre chlorophyll is *Bchl a* in both the green and brown species.

The following components are found in the reaction centres of *Rhodopseudomonas sphaeriodes, R.Virids* and *Rhodospirillum rubrum.*

1. Four molecules of *Bchl* a
2. Two molecules of Bacteriopheophytin (*Bph*)
3. 1 molecule of carotenoids
4. One or more quinones
5. One atom of non haeme iron

The reaction centres in most of the photosynthetic bacteria contain three chains polypeptides having a molecular rate of 20,000, 24,000 and 30,000. The *Bchl.* Bacteriopheophytin and Darotenoids are located on the first two polypeptides. In the primary photochemical reaction, the light quantum absorbed by the light harvesting pigments are transferred to the reaction centre (P870). In the reaction centre, the absorbed photons rais the chlorophyll to excited state and the electron is expelled from the molecule. The P870 now becomes positively charged. The charged P870 transfers the electrons to an intermediate electron acceptor which is reduced. This reduced acceptor then transfers its electron to another acceptor and ultimately the electron comes back to chlorophyll bringing the P870 to its normal state.

Cyclic photophosphorylation

The process of electron vibration here is the same as what has been explained above. When the electrons are moving from the ferrodoxin to cytochromes, a molecule of ATP is synthesized (ADP + Pi = ATP). This is called cyclic photophosphorylation, because the electron comes back to the chlorophyll in a circular fashion. The net turn of the cycle yields one molecule of ATP.

Non-cyclic photophosphorylation

Here the electron from ferrodoxin instead of moving down towards cytochromes is accepted by NAD, and the reaction is catalysed by ferrodoxin reductase. The reduced NAD now provides the necessary reducing power to fix the carbon dioxide.

The *Bchl* P890 now is in an excited state as it has donated its electron to ferrodoxin. In order to come to the normal state, it must receive an electron from some other source. The electron comes from the oxidation of organic acids, thiosulphates or H2S. These compounds get oxidised as a result of the elimination of an electron. The electron thus released moves via cytochromes back to *Bchl* P890. The chlorophyll molecule now comes back to the normal state and is ready to receive photons from light.

The scheme of photophosphorylation is called non-cyclic because the electron that comes back to *Bchl* is not from its own source but by the reduction of other compounds such as H_2S etc.

Light reaction in oxygenic photosynthesis

The light reaction or photochemical decomposition of water takes place in the grana region of the chloroplast. According to the current understanding, light reaction (Photochemical excitation) takes place in about 10^{-9} seconds. Within this short span of time, there will be the synthesis of molecule(s) of ATP and a molecule of NADPH. These two are utilized in the dark fixation of CO_2. In this section we will study the following aspects in detail with respect to light reaction.

(a) Source of oxygen released during light reaction
(b) Hill's reaction
(c) Arnon's work
(d) Quantum requirement
(e) Emerson effect and two pigment system
(f) Primary photochemical reaction (mechanism of pigment excitation)
(g) Non-cyclic Photophosphorylation.
(h) Cyclic Photophosphorylation
(i) Pseudocyclic photophosphorylation and
(j) Mechanism of ATP formation

SOURCE OF OXYGEN

The overall reaction of photosynthesis upto 1930 was believed to be the exact opposite of respiration and oxygen evolved during photosynthesis came from carbon dioxide.

$$6CO_2 + 6H_2O \text{——} C_6H_{12}O_6 + 6O_2$$

The discovery by Van Neil (1931), that in certain purple photosynthetic bacteria photosynthesis takes place in the presence of CO_2 but without releasing oxygen. Surely, if the source of O_2 were to be CO_2 even in the absence of water O_2 should have been released. These photosynthetic bacteria used H_2S as hydrogen donor instead of water. Accordint to Van Neil (1931), in purple bacteris H_2S, splits releasing H to reduce CO_2 and in the process elemental sulphur is being formed as follows :

$$CO_2 + 2H_2S \text{ ——— } (CH_2O) + H_2O + 2S$$

Van Neil (1931) further argued that, if elemental sulphur comes from H_2S in purple bacteria, in green plants which use water, O_2 should come from water only and not from CO_2. In order to account for the release of O_2 from water the equation should be modified as follows

$$6CO_2 + 12H_2O \text{ ——— } C_6H_{12}O_6 + 6H_2O$$

According to Van Neil (1931), water is split up by light (photolysis of water) into H^+ and OH^- ions. The H + ions are used to reduce CO_2, while the OH ions react with one another to form water and oxygen as follows :

$$4H_2O \text{ —— } 4H + 4(OH)^-$$
$$4(OH)^- \text{ —— } 2H_2O + O_2$$
$$CO_2 + 4H^+ \text{ —— } (CH_2O) + H_2O$$

Here (CH_2O) represents not the actual photosynthetic product but the empirical formula of a bigger molecule (glucose).

The work of Ruben, Kamen and Randall (1941), using radio active oxygen (O_{18}) clearly showed that O_2 came only from water and not from CO_2. When the experimental plant was supplied with labelled water (H_2O_{18}), the released oxygen was the O_{18} type, and when the plant was supplied with labelled CO_2 18, the O_2 released was the normal type (O_{16}).

HILL'S REACTION

Hill (1937), and Hill and Scaribrick (1940), showed that the initial reaction of photosynthesis can take place even in isolated chloroplasts in the cell-free environment. When isolated chloroplasts (obtained from ground leaves) are supplied with light, water and a suitable hydrogen acceptor, O_2 will be released and the H_2 acceptor gets reduced. A similar release of O_2 may be observed, when grana from disintegrated chloroplasts are suspended in water with ferric salts and benzoquinones acting as hydrogen acceptors. Besides these, chromates, indophenols etc. also can act as H_2 acceptors. The various H_2 acceptors used by Hill are called Hill oxidants. Hill attributed the release of oxygen to the splitting of water into H and OH by light. Hill's reaction proves the following :

(a) The entire photosynthetic activity including the reduction of CO_2 takes place within the chloroplasts, and

(b) O_{21} released during light reaction comes entirely from water.

ARNON'S WORK (production of assimilatory power)

The products of light reaction provide the necessary inputs for the reduction or assimilation of CO_2. That the assimilatory power (input) is utilized for the dark fixation of CO_2 was demonstrated clearly by the work of Arnon (1951).

According to Arnon (1951), the hydrogen released by water was accepted by coenzyme II (NADP). He found that in isolated chloroplasts, the release of O_2 takes place at a faster rate in the presence of NADP. Arnon also opined that ATP molecules will also be synthesized and he termed this as photophosphorylation.

$$ADP + Pi \xrightarrow[\textit{(inorganic phosphate)}]{\textit{energy}} ATP$$

Arnon also showed that CO_2 fixation takes place in the pigment free portion of the chloroplast (stroma), while the assimilatory power (ATP)/molecules and NADPH), is produced in the grana portion.

QUANTUMREQUIREMENT

The solar radiation that reaches the earth consists of two portions – the visible (400nm – 800nm), and UV, and infrared (less than 400 nm and more than 800 nm). Whether the light consists mainly of waves or particles, as far as photosynthesis is concerned, it is obsorbed in units of energy called photons or quanta.

The quantum yield may be explained on the basis of molecules of CO_2 reduced and amount of O_2 released. Consequently, a minimum quantum is the one required to fix one molecule CO_2 and release one molecule of O_2. According to Warburg and and Negelein (1922), the minimum quantum requirement is four i.e. 4 quanta are required to fix one mol of CO_2 and release one mol of O_2. This (requirement) concurs with the fact that every CO_2 molecule requires $2H_2O$ molecules for reduction i.e., 4 hydrogen atoms. Emerson and his co-workers (1938), however opine that eight light quanta is the minimum requirement. This has been further confirmed by the work of Govindjee (1968), on *Chlorella* plants. Thus two quanta of light are required for the transfer of every atom of hydrogen during photosynthesis.

EMERSON EFFECT AND TWO PIGMENT SYSTEMS

Emerson and Lewis (1943), worked on different wavelengths of light (monochromatic light) as to their quantum yield and came to the conclusion that 8 quanta of light are required for the reduction of one molecule of CO_2 (or to release one molecule of O_2). Accordingly, the quantum yield is 1/8 or 12%. Since 4 electrons are required in the release of one molecule O_2, Emerson opined that 2 light quanta are required to excite one electron. Proceeding further, Emerson *et al* (1943), determined the quantum yield of different beams of monochromatic light and found that the average quantum yield (12%) suddenly dropped near the far red end of the spectrum. This decline in quantum yield is called red drop. Red drop begins at wavelengths greater than 680 nm in green plants and greater than 650nm in red algae. The phenomenon of red drop was also confirmed by the works of Haxo and Blinks (1950), on the red alga *Porphyra neregtis*.

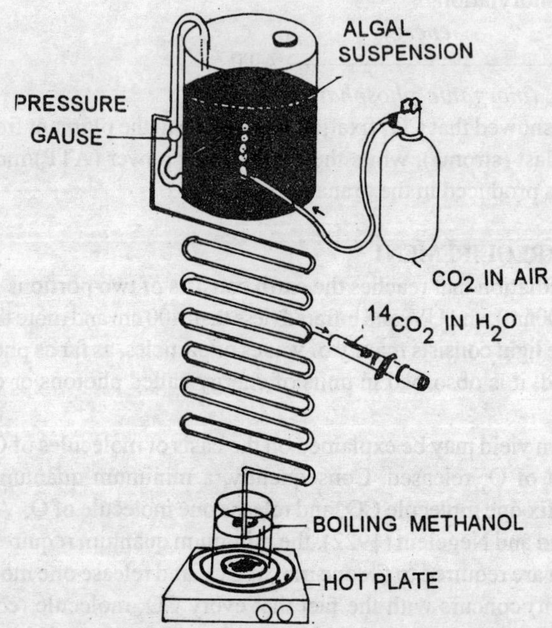

Fig. 5.19 Photosynthesis
Apparatus used by Calvin et al for short time exposure to C^{14}

The phenomenon of red drop was studied further by Emerson and Chalmers (1958), who found out that the reduction in the rate of photosynthesis due to red drop can be rectified by providing a beam of monochromatic light of shorter wavelength. In other words, by providing shorter wavelengths of light either simultaneously or quickly alternating with a longer wavelength of light, the rate of photosynthesis can be enhanced (effect of red drop is reversed).

This enhancement of photosynthetic rate under the influence of two wavelengths (long and short) of light is called Emerson effect. This may be represented as follows :

$$E = \frac{\text{Rate of evolution of } O_2 \text{ in double beam - Rate of Evolution } O_2 \text{ in red beam (short)}}{\text{Rate of evolution of } O_2 \text{ in far red beam (long)}}$$

Reason for Emerson effect

The discovery of the Emerson effect led to the idea that there must be two different pigment systems in the plants which differed in their absorption peak (of light). While one pigment system was more efficient in longer wavelength, the other was efficient in shorter wavelength.

Obviously, under a specific wavelength (either short or long), the rate of photosynthesis will not be maximum as one or the other pigment, system will be ineffective. Butler (1966), discovered two forms of chlorophyll viz *chl a* 673 and *chl a* 683 (with absorption peaks at 673nm and 683 nm respectively). Clayton (1966), discovered another form of *chl a* with absorption peak at 700nm which he named P700.

It has now been confirmed by various researches that two pigment system PS I and PS II are involved in photosynthesis, each pigment system having different absorption peaks and consisting of *chl a* and other accessory pigments. The two pigment systems may be removed from the thylakoids, by treating the chloroplast with mild detergents. These can be separated on polyacrylamide gel when they appear as discreet green bands. The details of the two pigments system are as follows :

According to Malkin (1982), Anderson (1982), and Barber (1983), PS I consists of Chl *a* 670, Chl *a* 680 and Chl *a* 710 (P 700), Chl *b* and some β cartoenes. P700 is the reaction centre in PS I into which light energy is transferred by all other pigments. Besides, PS I also consists of iron containing proteins called Fe-S proteins and Cytochromes. PSI is weakly fluorescent.

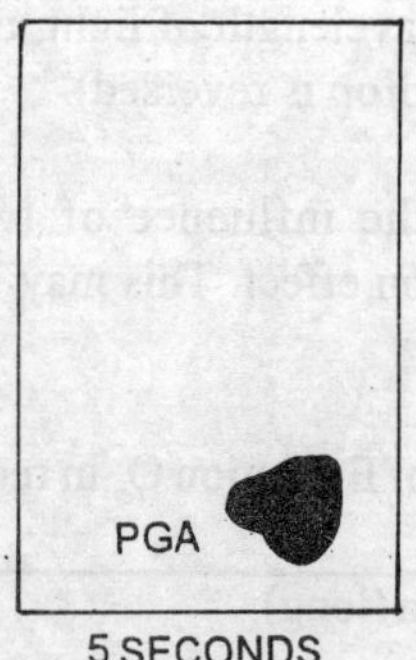

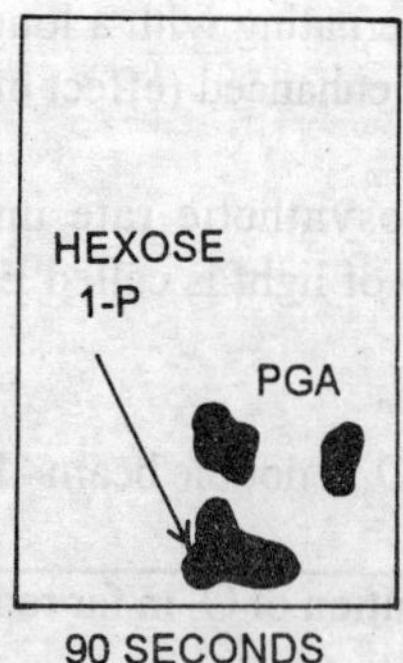

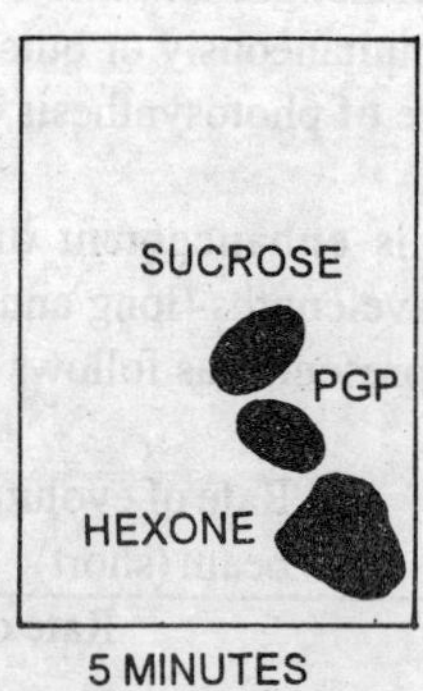

Fig. 5.20 Photosynthesis

Auto radiograph (Diagrammatic representation of chromatograms showing the fixation of C^{14} after different periods of exposure to light

Pigment System II (PSSII) :

It has several forms of Chl *a* with P680 (Chl a 680) being the reaction centre. A little amount of Chl *b* and β carotenes are also present. There is also a colourless Chl *a* called Pheophyrin (Klima and Krasnovskii, 1981), which acts as the primary electron acceptor together with these are present quiones (Q) and manganese proteins.

Light Harvesting complexes

Electrophoretic separation of pigments from Chloroplasts, reveals the presence of two other green bands besides PS I and PS II (Salisbury and Ross, 1986). Each band contains Chl *a* + Chl *b* but very little β carotene. All these pigments are protein bound. According to Salisbury and Rosss (1986), these green bands represent light harvesting complexes of pigments and proteins; one band functions in association with PS I and the other with PS II. The function of the light harvestng complexes is to absorb light energy and to transfer it to the appropriate pigment system.

Co-ordination between PS I and PS II : According to Junge (1982), each granum has about 200 units of PS I and II, while the number is variable in stroma thylakoids. It was originally believed that chloroplasts contain equal quantities of PSI and PS II, but according to Melis and Brown (1980), the ratio of distribution of PS I and PS II varies with the species and other environmental conditions. Anderson and Melis (1983), believe that grana consists mainly PS II, while stroma thylakoids have mainly PS I.

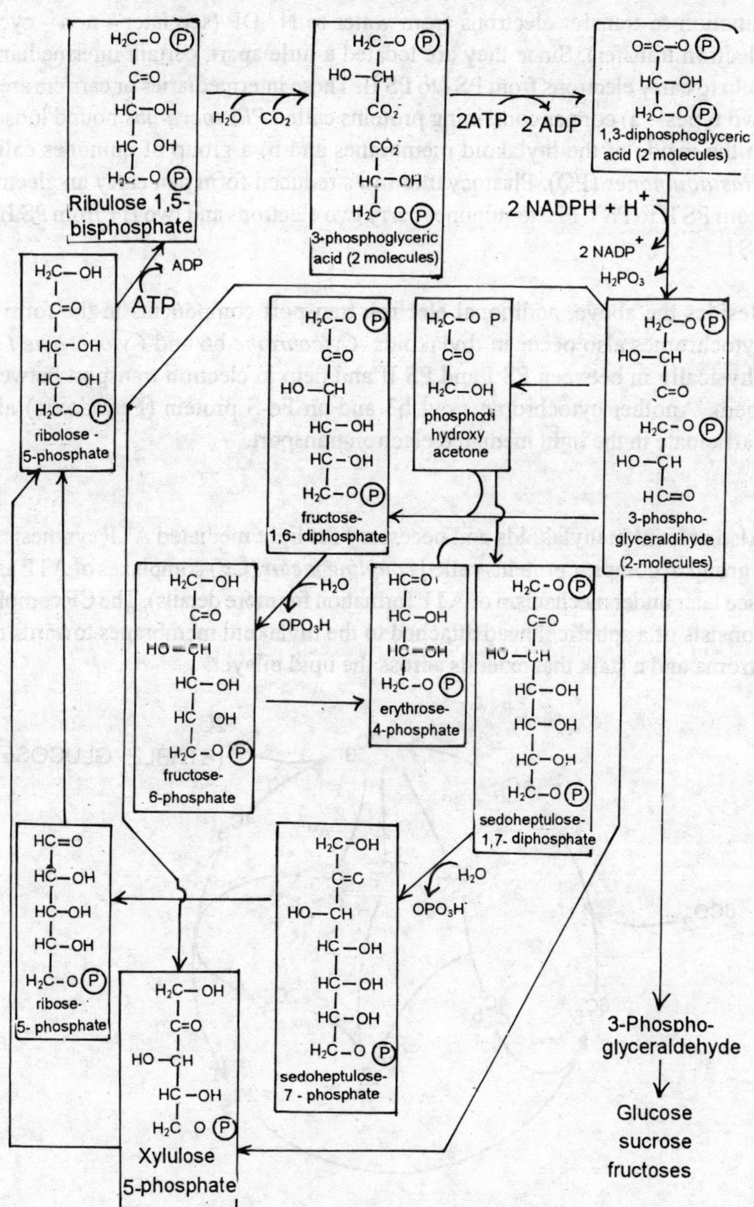

Fig. 5.21 Photosynthesis

Systematic representation of C3 pathway (Calvin Cycle

PS I and PS II should function in close co-ordination because they jointly function to transfer electrons from water to NADP (see later - non - cyclic electron transfer). Since they are located a little apart, certain intermediaries help to carry electrons from PS I to PS II. These intermediaries or carriers are of two types – a) copper containing proteins called *Plastocyanins* bound loosely to the inside of the thylakoid membranes and b) a group of quinones called *Plastoquinones* (PQ). Plastocyanin in its reduced form can carry an electron from PS II to PS I. Plastoquinones carry two electrons and two H+ from PS II to PS I.

Besides the above, additional electron transport components in the form of cytochromes also occur in thylakoids. *Cytochrome* b6 and *Cytochrome f* lie physically in between PS I and PS II and help in electron transport between them. Another cytochrome - cyt b3 and an Fe-S protein (Ferredoxin) also participate in the light mediated electron transport.

Also present in thylakoids and necessary for light mediated ATP synthesis is, a group of complex proteins called *coupling factor* (CF) complexes of ATP ases (see later under mechanism of ATP formation for more details). The CF complex consists of a spherical head attached to the thylakoid membranes towards the stroma and a stalk that extends across the lipid bilayer.

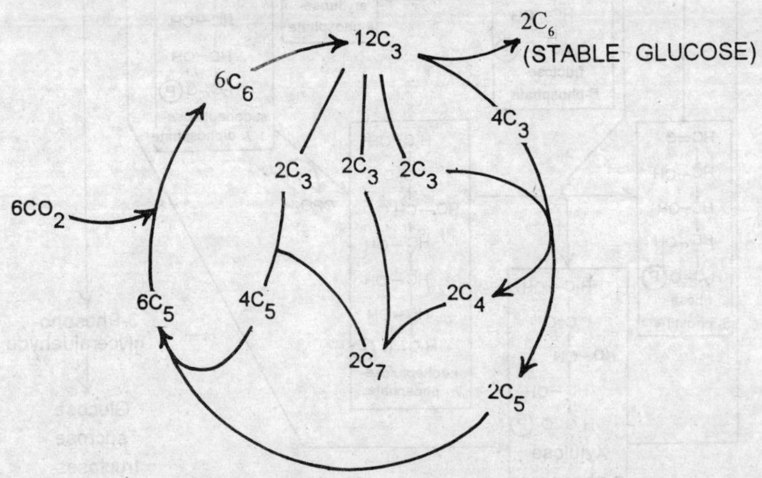

Fig. 5.22 Photosynthesis
Balance Sheet of carbon Calvin's Cycle

MECHANISM OF PIGMENT EXCITATION

Normally, the chlorophyll molecule wil be in its stable state (no excitation), in the sense the two electrons of its atoms will be spinning in opposite directions and the molecule has no magnetic movement.

When the chlorophyll molecule absorbs light energy in the form of photon one of the electrons is raised to a higher orbit (higher energy). However, the spin of the electrons continues to be in opposite directions, and hence to magnetic movement. The unexcited state of the chlorophyll molecule (before the absorption of light) is called the *ground state* or the *single state.* Here, the energy level is zero, hence it is also called the E_0 state. After the excitation, the stage is known as the first excited state or the S_1 state, with the energy being E_1. The life span of the S_1 state is very short and very soon (10^{-9} sec), the excited electron returns to the ground state by losing energy in the form of heat, light, or by the transfer of energy. The chlorophyll molecule reaches the S_1 state by absorbing the light at the red level (660 nm). If the light is absorbed at the blue level (400 nm), the chlorophyll molecule isknown to reach the S_2 level (with energy level being E_2), because the 'blue photon' has more energy. However, even here the magnetic movement is nil as the spin of the electrons is still in the opposite directions. The electron of S_2 state reaches the ground state via the S_1 state by losing energy either as heat or as heat of light (fluorescence or phosphorescence). Since the electron reaches the ground state very soon (10^{-8} or 10^{-9} see), the energy cannot be put into use (it canot be converted into chemical form).

Under certain circumstances, an electron at the S_1 state may lose some energy in reversing the spin and drops into a low orbit. At this state, there will be two electrons in two different orbits, but spinning in the same direction. The molecule is now said to possess magnetic movement with electrons oriented in three different ways relative to the axis. This is called the *triplet state* or the *metastable state* (T_1 state). The conversion of S_1 state electron to T_1 state is called *Inter system crossing.*

The lifespan of T_1 state electron is longer (10^{-2} or 10^{-3} sec). The electrons do not dissipate energy quickly, for they have to reverse their spin before reaching the ground state. Since the electron in the T_1 state remains energised for a long time, the energy can be made use of for chemical reaction (conversion into chemical energy). The electron at the T_1 state is dislodged from the chlorophyll molecule and is shuttled between a series of electron acceptors, the energy is utilized to produce ATP and (reduced) NADPH molecules. The series of electron acceptors constitute the electron transport system (ETS).

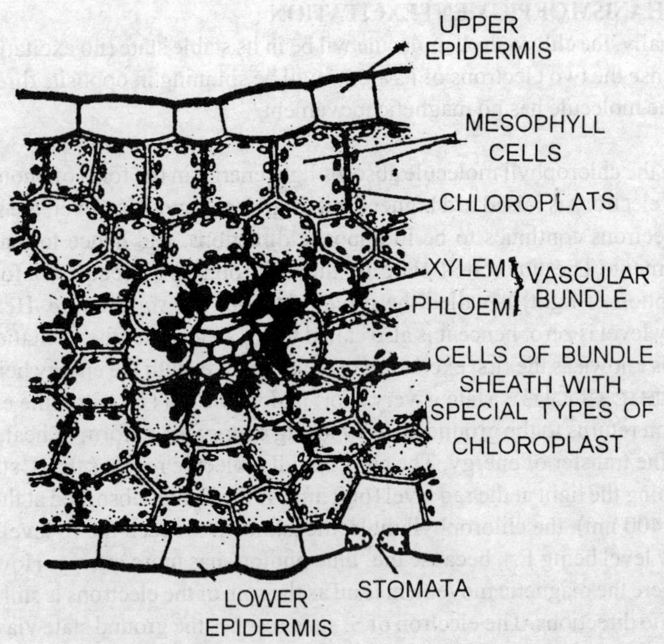

Fig. 5.23 Photosynthesis

TS of a monocat leaf (grass) showing Kranz anatomy

Photophosphorylation

Arnon and Frankel (1954), showed that isolated chloroplasts, under illumination can produce ATP molecules. The reaction is believed to take place as follows.

$$ADP + Pi \xrightarrow[\text{Chlorophyll}]{\text{Light}} ATP$$

(ADP = Adenosine Diphosphate)

ATP = Adenosine Diphosphate)

Pi = Inorganic phosphate

Aron *et al* (1954), named this phenomenon as Photophosphorylation (addition of phosphate group to ADP under the influence of light energy). These ATP molecules are utilized to reduce PGA to PGAL during the fixation of CO_2. They are also required to phosophorylate Ribulose 5 phosphate to Ribulose 1.5 diphosphate (For details, see dark reaction). The following types of photophosphorylation have been discovered in plants.

1. Cyclic Photophosphorylation

2. Non - Cyclic Photophosphorylation, and
3. Pseudocyclic Photophosphorylation

Cyclic Photophosphorylation

Here only PS I participates. The light energy absorbed by the light have harvesting complex which is transfered by inductive resonance to P700. P700 can accept light energy only of a wavelength longer than 680 nm. The excited electron of PQ700 is first accepted by an unknown acceptor (x), then to ferredoxin. This causes ferredoxin to release one of its own electrons (energy rich), which shuttles through the chain consisting of Cyt b6, Cyt f. state and now ready to receive one more unit of radiant energy.

The electron released from the chlorophyll (P700) is returned in a continuous chain i.e., it travels in cyclic fashion, hence the name, cyclic electron transport or cyclic photophosphorylation.

Synthesis of ATP during cyclic photophosphorylation

The electron ejected from P700 has an energy of 0.42 v which gets reduced to 0.417 when it reaches ferredoxin. Here the electrons passes through 3 down hill migration (Cyt b6, Cyt f and Pi) steps before it is returned to P700. The potential gap between Ferrodoxin to P700 (ground state) is about one volt which is sufficient to produce at least two molecules of ATP. ATP molecules are formed at two stages – a) between ferredoxin and Cyt b6 and b) between Cyt b6 and Cyt f

In cyclic photophosphorylation, water does not participate; hence there is neither evolution of O_2 nor is there any formation of reduced NADPH molecule (for there is no hydrogen donor). Further, PS II is not involved in any stage.

According to Park and Sane (1970), there are two types of PS I, PS Is and *PS1g*. *PS 1s* is present in the stroma thylakoids and *PS 1g* is present in the granum. *PS 1s* participates in cyclic photophosphorylation, while *PS 1g* participate in non-cyclic photophosphorylation together with PS II.

Ramreiz et al (1968), opine that cyclic photophosphorylation is of limited occurrence in plants, and a high rate of cyclic reaction may even retard the CO_2 fixation as NADPH will not be available. They argue that cyclic reaction is not involved in the main path of carbohydrate synthesis as it can only provide part of ATP requirement for CO_2 fixation.

Non-Cyclic Photophosphorylation

This is the major pathway of light reaction involving both PS I and PS II. Non-cyclic photophosphorylation is also called the 'Z' scheme electron transport because of the zig-zag fashion of electron travel. It takes place as follows :

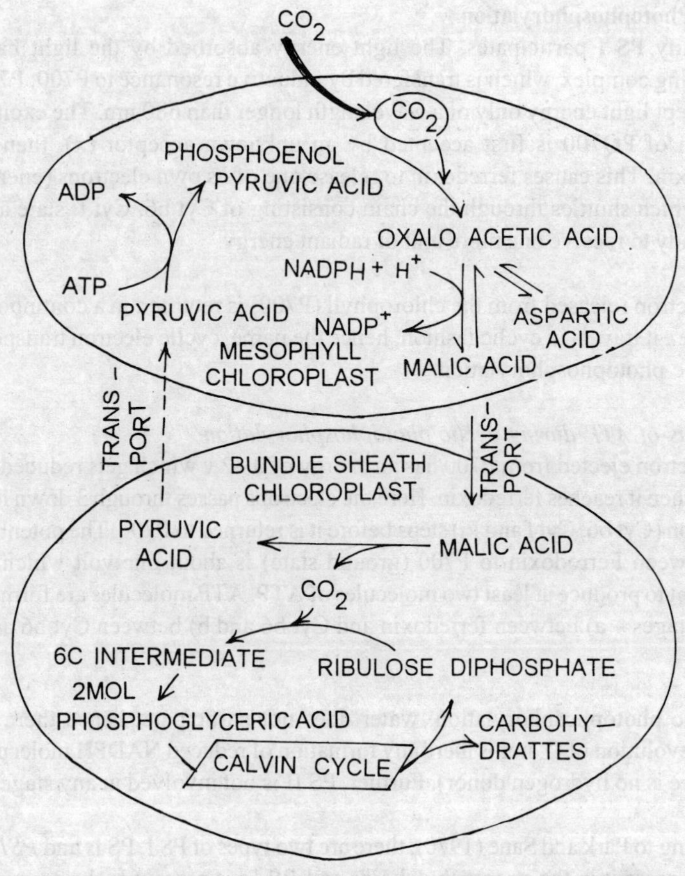

Fig. 5.24 Photosynthesis
Systematic representation of Hatch and Slack pathway

(a) The radiant energy is trapped by PS I (P700), which ejects one if its electrons and is trapped by X, FRS, and finally goes to ferredoxin. The reduced ferredoxin now transfers its electrons to NADPH. NADPH gets reduced. PS I, however remains in the excited state as it has not got back its electron.

(b) Meanwhile, PS II absorbs light energy (680nm) and ejects one of its own electrons which is trapped by an unknown quinone acceptor (Q). From Q–the electron moves downhill to cyt, b6, plastoquinine, cyt f, Plastocyanin (PC), and finally to PS I. PS I which is in an excited state, now returns to the ground state as it has got an electron from PS II.

How does PS II come back to the ground state now that it has sent its electron to fill up the 'hole' in PS I?. Because, unless the PS II returns to the normal state, the system cannot contunue functioning. The answer lies in the dissociation of water molcculcs. Water splits into H^+ and OH^+ ions. What is the mechanism of this splitting of water? Some physiologists call it photolysis of water, others, the breakdown of water. In any case, the excited PS II returns to the normal state by getting an electron from the OH^- ions of water. According to Salisbury and Ross (1986), the excited PS II returns to the ground state by attracting an electron from an adjacent Mn^- protein, which in turn gets an electron from the OH^- of water. The H^+ ions released by water are accepted by NADPH to become NADPH + H^+. Thus, a reduced NADPH is formed.

The electron transport system or photophosphorylation is called non-cyclic because the excited PS I returns to the normal state by getting an electron from PS II, while the excited PS II returns to the ground state by receiving an electron from water. Hence, the movement of electron is not in a cyclic fashion.

ATP Synthesis during non-cyclic photophosphorylation

The electron (+ 0.8 v) released by PS II is accepted by a quinone from where the downhill migration begins to reach PS I via Cyt b6, Cyt f and PC. One molecule of ATP is synthesized when the electron shuttles between PQ and Cyt f.

Release of oxygen and formation of NADPH + H^+

Four molecules of water split up for every turn of non-cyclic photophosphorylation. This is necessary (see fig) to provide four electrons to the excited PS II to bring it back to the ground state. Similarly, four H + ions are required to reduce 2 molecules of NADPH.

These reactions take place as follows :

$$4H_2O \longrightarrow 4H^+ + OH|^-$$

$$4OH^- \longrightarrow 2H_2O + O_2$$

(4 electrons from OH reach the excited PS II)

$$2NADPH + 4H^+ \longrightarrow 2nadph + 2H^+$$

(four electrons are required to reduce 2 molecules of NADPH and these come) from the excited PS II via ferredoxin)

Thus the end products of non-cyclic photophosphorylation are :

(a) One molecule of ATP

(b) Two molecules of reduced NADPH

(c) 2 molecules of water and

(d) One molecule of O_2.
While a and b are used for the dark fixation of CO_2, c and d are byproducts.

Comparison between cyclic and non-cyclic photophosphorylation

	Cyclic	Non-cyclic
1.	Only PS I participates	PS I and PS IIparticipatre
2.	PS I gets back the electron in a cycle fashion	PS I gets back the electron from PS II from water
3.	Only ATP molecules are formed	Both ATP and NADPH molecules
4.	Water does not participate, hence no evolution of O_2.	O_2 evolved as water participates
5.	Found predominantly in bacteria	Found predominantly in green plants.

Pseudocyclic Photophosphorylation
Arnon and his co-workers demonstrated another kind of Photosphosphorylation, when illuminated chloroplasts produced ATP molecules in the absence of CO_2 and NADP, but in the presence of FMN or Vit K and oxygen. This involves the reduction of FMN by water with the evolution of oxygen. The reduced FMN is reoxidizable by oxygen. Hence, Arnon (1954), called it oxygen dependent FMN catalysed photophosphorylation or pseudocyclic photophosphoyrlation.

MECHANISM OF ATP FORMATION

The source of energy for the synthesis of ATP molecules from ADP comes obviously from the electron transport chain. There are many evidences which point out to this relationship brought about by some common agents. For instance, if some uncoupling agents are introduced into the system, electron transport continues, while the ATP synthesis is inhibited. If the uncoupling agent is removed, ATP synthesis is resumed. Inhibition of electron transport also inhibiting ATP synthesis is another evidence to show that both are coupled.

While there does not seem to be any doubt about the coupling of electron flow and phosphorylation, the mechanism is as yet not fully understood (according to some physiologists). The following are the three hypotheses proposed for the mechanism of ATP synthesis.
(a) Conformational coupling

(b) Chemical coupling
(c) Chemiosomotic coupling

(The mechanism of ATP synthesis is the same for both photophosphorylation during photosynthesis as well as for oxidative phosphorylation during respiration.)

Conformational coupling

During electron transport, both in mitochondria as well as in chloroplasts, some conformational (structural) changes occur which facilitate the release of energy resulting in the production of ATPase mediated ATP synthesis, ATPase is a reversible action enzyme which decomposes ATP into ADP and Pi under less energy conditions, while facilitates synthesis of ATP during high energy conditions. According to Devlin and Witham (1983), electron micrographs of membranes indicate structural variations during peak activity. But there does not seem to be any definite relationship between activated membrane and ATP synthesis.

Chemical coupling

This hypothesis believes in the presence of a coupling factor (most probably a protein), mediating the energy transfer between electron transport and ATP synthesis. Initially, the coupling factor (CF) forms a high energy CF complex with one of the electron carriers. Later, an inorganic phosphate replaces the electron carrier in the CF complex. This results in the formation of high energy rich phosphate to ADP resulting in the formation of ATP.

Chemiosmotic coupling

Among the hypothesis proposed for the mechanism of ATP synthesis, this is by far the most acceptable, Originally proposed by Michell during 1960s for oxidative phosphorylation in mitochondria, it has recently been applied to photophosphorylation in chloroplasts also (Jagendorf, 1975).

As it relates to oxidative phosphorylation, the flow of electrons from NADH to atmospheric oxygen through the respiratory electron transport, influences the extrusion of protons from the mitochondrial matrix. As a result, a proton concentration gradient is produced across the mitochondrial membrane. In effect, the energy of electron flow in the respiratory chain is partially conserved as the *proton concentration gradient* or *proton motive force* across the inner mitochondrial membrane. The protons return to the mitochondrial matrix down the electrochemical gradient through ATPase resulting in the formation of ATP from ADP and Pi.

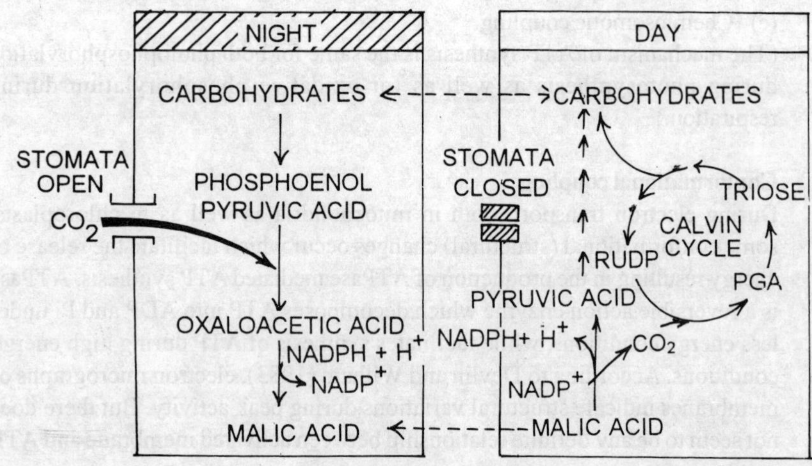

Fig. 5.25 Photosynthesis

Systematic representation of CO_2 fixation in CAM Plants

Ubiquinone, one of the components of the respiratory chain known to transfer protons across the mitochondrial membrane through alternate oxidation and reduction. The flow of protons across the membrane is due to a force called *Proticity* (Noggle and Fritz, 1983), which is comparable to electricity. According to Noggle and Fritz (1983), the word chemiosmotic was proposed to emphasize the fact that chemical energy (oxidation energy) is converted to osmotic energy (difference in the proton concentration gradient between the inner and outer faces of the membrane).

Oxidative phosphorylation may be halted by treating the membrane with uncouplers. These are weak acids which dissolve in the protoplasmic membrane and disturb the proton concentration gradient by allowing the free flow of protons. Thus the proton uncouplers (protonophores) disturb the ATP synthesis.

The Chemiosmotic theory also explains ATP formation in chloroplats. Jagendorf (1975), demonstrated that a pH gradient established across the thylakoid membranes stimulates the ATP synthesis. The electrons are transported across the membrane to the interior through the electron transport chain. Light energy provides the necessary drive the flow of electrons from water to NADP and for transport of protons across the membrane. Plastoquinone, an electron acceptor transfers electrons to Cyt f and picks up H + ions on the outside of the thylakoid membrane and releases protons to the interior of the channel. This results in the building up of a pH gradient (the interior of the thylakoid becomes acidic due to accumulation of protons) across the membrane. The accumulated protons flow from the inside to the stroma side of the membrane through pathways of coupling factor (which include stalks and knobs), which are the sites of photophosphorylation. The downhill migration of protons provides the necessary energy for the addition of Pi to ADP to form ATP, with ATPase acting as a mediator.

DARK REACTION OF PHOTOSYNTHESIS (carbon assimilation)

The source of carbon for reduction is atmospheric CO_2 for green plants. For photosynthetic bacteria, carbon may come from Co_2 or from organic compounds. The process of CO_2 assimilation in photosynthetic bacteria is of the following types :

1. Reductive pentose pathway or Calvin cycle
2. Reductive carbosylic acid cycle or pyruvate synthetase pathway
3. Organic carbon assimialation pathway

CALVIN CYCLE (reductive pentose pathway)

Calvin cycle or the C_3 cycle, wich

is the predominant pathway of CO_2 fixation in higher plants is seen in photosynthetic bacteria also. Researches carried out on *Rhodospirillum ruburm* and *Thiobacillus* spp using radio isotopes of carbon (Fog, 1968), have confirmed this. Sugar phosphates are, however, less as these bacteria are unable to utilize or store them.

The phenomenon of thermochemical reduction of CO_2 was first established by Blackmann in 1905. He opined that light is not necessary for the reduction of CO_2. In dark reaction the products of light reaction (ATP and NADPH + H) are used to reduce CO_2 into glucose. But the precise manner in which CO_2 is absorbed, and reduced resulting in the formation of a molecule of glucose was understood fairly recently, thanks to the availability of radioactive isotopes. Utilizing the isotopes of carbon dioxide, one can trace the different stages of entry of

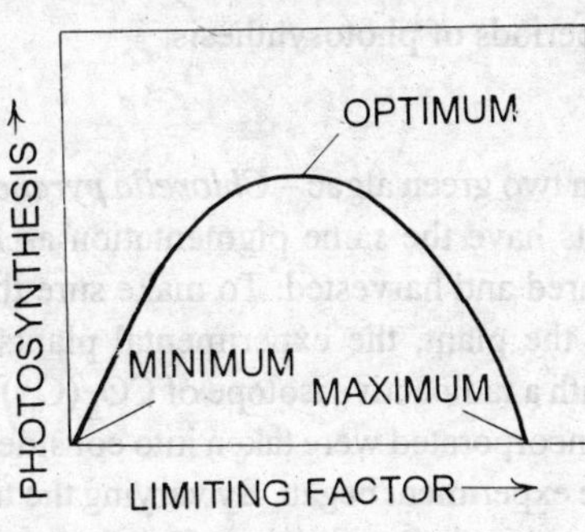

Fig. 5.26 Photosynthesis
Sach's concept of 3 cardinal points

CO_2 into the plant system untill it gets converted into glucose.

Many physiologists, biochemists etc. have contributed a great lot towards the understanding of the path of carbon in Photosynthesis. The names of Bassham (1957 - 1964), Melvin Calvin (1961), Benson (1951) etc are worth mentioning in connection with the untravelling of the secrets of CO_2 fixation. Melvin Calvin was awarded the Nobel Prize in 1961 for tracing the CO_2 reduction cycle, which is also known as the Calvin cycle.

In CO_2 reduction cycle, what we have to precisely understand is how the CO_2 enters the plant system, which is the compound that accepts the CO_2 molecule and how a molecule of glucose is synthesized and how the compound (that originally accepts the CO_2) is regenerated. In addition, the system should explain the acceptance of 6 CO_2 molecules, their reduction and formation of only one molecule of glucose for every turn of the cycle.

CALVIN CYCLE

This is the major pathway for the fixation of CO_2 in green plants and in some bacteria. It is called Calvin Cycle owing to its discovery by Melvin Calvin of the university of California, USA. It is also called C_3 pathway because, the first stable compound formed after the entry of CO_2 into the plant system is a triose sugar (C_3) viz phosphoglyceric acid (PGA).

Some experimental details of Calvin's work

In order to understand the significance of the discovery of CO_2 reduction cycle, we should understand the experimental methods followed by Calvin.

To trace the various steps of conversion of CO_2 after its entry into the plant system until the formation of glucose, Calvin allowed the experimental plants to photosynthesize for various periods of time and analysed the various compounds formed after different periods of photosynthesis.

Clavin chose for his experimentation two green algae – *Chlorella pyrenoidosa* and *Senedesmus olbiqus*. The plants have the same pigmentation as higher green plants and can be easily cultured and harvested. To make sure that the compounds are already present in the plant, the experimental plants were starved of CO_2, and later supplied with a radioactive isotope of CO_2 (C_{14}). Only those compounds in which C_{14} was incorporated were taken into consideration because they were produced after the experiment began. By varying the time of photosynthesis (1/16 sec to 5 sec), and then killing the plant to stop photosynthesis and analysing the intermediate compounds formed, Calvin and his co-workers were able to map the whole of the CO_2 reduction cycle.

Two, very simple but efficient tools is namely *Chromatography* and *Radioauthgraphy,* to trace the path of carbon in photosynthesis.

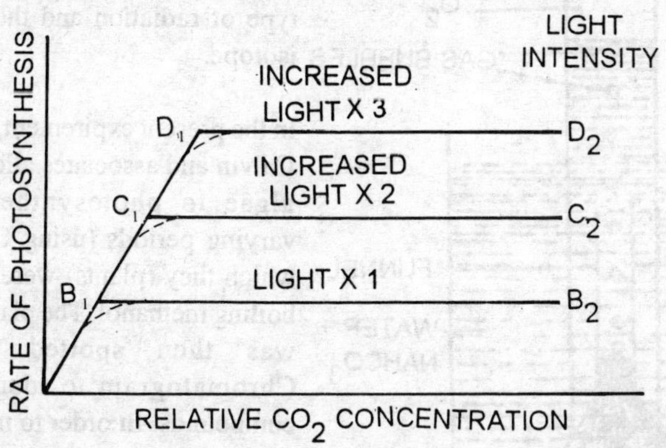

Fig. 5.27 Photosynthesis

Graph representing Law of Limiting Factors

Chromatography, in particular, paper chromatography is an ideal technique for the separation of compounds present in a small mixture. The extract (mixture) of the plant to be analysed is placed as a spot on paper and a solvent is allowed to run. This solvent due to capillary action carries with it the different substances in the extract. The rate of migration of the differnt substances depends upon their solubility and specific gravity.

The filter papter (Chromatogram) should be removed from the solvent before it reaches the edge of the paper and dried. By spraying a suitable reagent (For e.g. ninhydrin for amino acids), the compounds can be detected as several spots widely separated on the paper. The substances may be identified by their relative positions on the Chromatogram or by comparing with a standard Chromatogram. (In a standard Chromatogram, compounds whose identity are known are spotted, and their position ismarked after the solvent run).

Radioautography is a technique to find out, if a particular source consists of a particular type of isotope. Different radioactive isotopes emit different types of radiation. Chiefly, there are 3 types. These are : (a) Alpha rays (b) Beta rays (c) Gamma rays.

When we use a particular we already know what is the type of emission. For instance, C_{14} emits β rays. When the rays are directed to a photographic film they from characteristic markings, and by these one can identify the type of radiation and thereby the isotope.

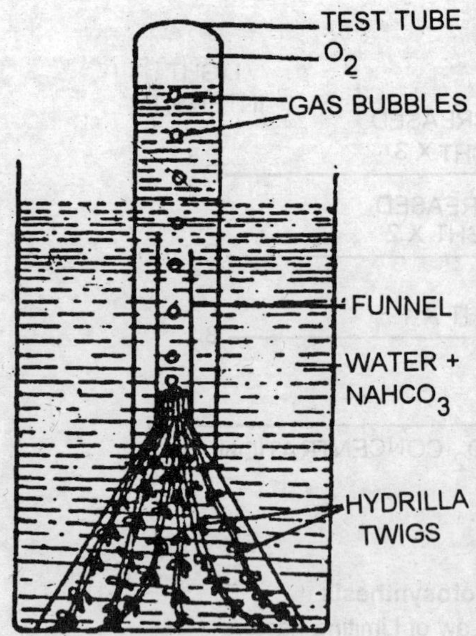

Fig. 5.28 Photosynthesis
Experiment to demonstrate the release of O_2 during photosynthesis

In the present expirement, Melvin Calvin and associates allowed the algae to photosynthesise for varying periods (using C_{14}), after which they (plants) were killed in boiling methanol. The plant extract was then spotted on the Chromatogram to identify the compounds. In order to make sure that these contain C_{14}, the chromatogram was held against a photographic film of equal size. Only those compounds which had incorporated C_{14} leave markings on the film and these were taken into consideration (as produced during experimentation). In order to vary the time of photosynthesis, flashes of light from 1/16 sec to 5 sec were given to identify the earliest compound formed. In this way, Calvin and his associates found out all the intermediaries in the CO_2 reduction cycle.

The first stable compound formed after the initial entry of CO_2 is a triose (PGA). The triose cannot be the acceptor molecule for it has to initially, have a 2C compound to accept one CO_2 molecule to become triose. Benson, however, soon found out that a 5 carbon compound is the first acceptor molecule (Ribulose 1.5 diphosphate). After accepting a molecule of CO_2, an unstable 6C compound is formed, which immediately breaks up into 2 molecules of PGA (triose), the first stable compound.

The essentials of the Calvin cycle are as follows. The cycle should be self exclusive in that it should remain unaltered, but provide for the continuous entry of 6 molecules of CO_2, formation of one molecule of glucose and regeneration of six molecules of Ribulose 1.5 diphosphate (RUDP) 1, the initial acceptor molecule. The important events of CO_2 cycle can be detailed as under.

(a) Fixation of CO_2
(b) Reduction of PGA
(c) Formation of sugars, and
(d) Regeneration of RUDP

Fixation of CO_2

CO_2 is first acccpted by molecules of a pentose (5C) sugar Ribulose 1.5 diphosphate, and an unstable 6 carbon compound is formed. This compound (tentatively called carboxy dimsutase system) immediately breaks up into 2 molecules of a triose (3C) called phosphoglyceric acid (PGA). The reaction is catalyzed by the enzyme carboxy dismutase of Rubp carboxylase (Rubisco). For every molecule of CO_2, 2 mols of PGA are formed; hence 12 PGA moles for $6CO_2$ mols.

Reduction of PGA

The molecules of PGA are reduced to phosphoglyceraldehyde (PGAI) by NADPH + H. The reaction requires energy and hence. ATP participates. The enzyme triose phosphate dehydrogenase catalyses this reaction.

Rudp + CO_2 H_2O ——— PGA (2 molecules)

ATP

PGA + NADPH + H ——— PGAI

The above reaction takes place at two stages. In the first stage, PGA + ATP Phosphoglycerokinase 1,3 - diphosphoglyceric acid + ADP In the ssecoud stage.

1.3 Diphosphoglyceric acid + NADPH + H —— 3 PGAI + NADP

Since 6 molecules of CO_2 are taken into the system (by 4 mols of Rudp), 12 PGAL mols are formed after reduction of the 12 molecules of PGA. Only two are utilized for the formation of sugars, while the rest are channeled back for the regeneration of Rudp molecules. Several intermediate compounds are formed in the process.

Formation of sugars

The PGA molecule (one out of two) undergoes isomerization to form a mol of dihydroxyacetone phosphate. The enzyme triose phosphate isomerase catalyses this reaction. Ultimately, sugars are formed as follows :

Triosephosphate

1. 3 PGAL ——— Dihydroxyacetone phosphate

Isomerase

2. 3 PGAL + Dihydroxyacetonephosphate

Aldolase

——— Fructose 1.6-diphosphate

Fructose 1,6-diphosphate now gets converted into fructose 5 monophosphate due to the loss of one phosphate group under the influence of the enzyme phosphotase.

3. Fructose 1.6-dip $\xrightarrow[+H_2O]{\text{Phosphatase}}$ Fructose 6. Monophosphate (FMP).

Fructose 6 monophosphate isomerises into glucose monophosphate under the influence of the enzyme isomerase.

4. EMP Isomerase Glucose 6 monophosphate (GM)

Now FMP or GMP may undergo dephosphorylation to form either fructose or glucose. These monosaccharides unite to form sucrose and various other carbohydrates.

Regenration of Rudp

In the above explanation of formation of sugars, it has been pointed out that Rudp accepts a molecule of CO_2 to ultimately form a molecule of sugar. But, the overall formula for photosynthesis accounts for the entry of 6 molecules of CO_2, hence, there must be 6 molecules of Rudp to accept the CO_2 molecules and for every turn of the cycle, this should result in the formation of 6 molecules but there is only one molecule of glucose. Then what happens to the remaining 6 molecules of hexose (5moles of 6C). Obviously they undergo transformation to regenerate 6 molecules of Rudp. This is necessary format Rudp to be ready to accept and be self-sufficient; and there must always be 6 molecules of Rudp ready, to accept 6 molecules of CO_2 and form 6 molecules of glucose, the cycle will cease to operate for the next turn of the cycle there will be no Rudp to accept CO_2 molecules. Thus for the entry of every 6 CO_2 molecules, the output is one mol of hexose and the rest go for the regeneration of Rudp. The balance sheet of the molecules of carbon from the entry of CO_2 to the generation of Rudp is given in Fig. In the following equations, regeneration of one mol of Rudp is explained (and this may be multiplied to obtain 6 molecules of Rudp).

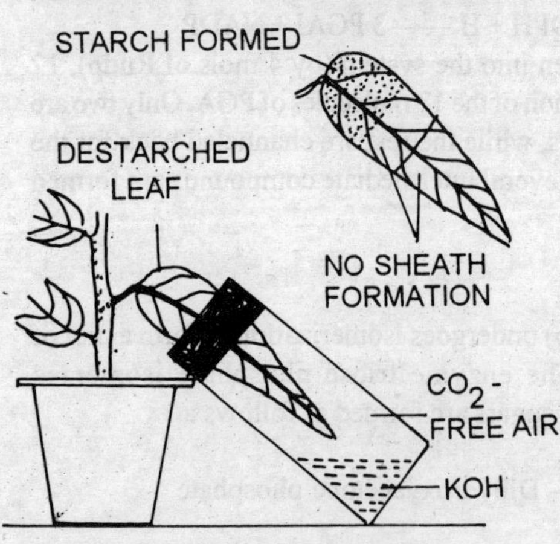

Fig. 5.29 Photosynthesis
Experiment to demonstrate the necessity of light for photosynthesis

1. Fructose 6 Phosphate + PGAL $\xrightarrow{\text{Transketolase}}$ Erythrose 4, Phosphate+ Xylulose 5. Phosphate
2. Erythrose 4, Phosphate + Dihyrosxyacetone phosphate $\xrightarrow{\text{Aldolase}}$ Sedophetulose 1.7. Diphosphate
3. Sedoheptulose 7. Phosphate + PGAL $\xrightarrow{\text{Phosphatase}}$ Xylulose 5 H_3PO_4.
4. Sedoheptulose 7. Phosphate + PGAL $\xrightarrow{\text{Transketolase}}$ Xylulose 5 Phosphate + Ribose 5 Phosphate

Xylulose 5 Phosphate and Ribose 5 phosphate get converted into Ribulose 5 phosphate under the influence of the enzyme phosphophentose isomerase

5. Ribulose 5 Phosphate $\xrightarrow[\text{-Kinase}]{\text{Phosphopento}}$ Rudp

Reductive carboxylic acid cycle

This has been demonstrated in green bacteria such as *Chlorobium thiosulfatophilum* and *Chromatium* spp (Evans et albugo, 1966). In this type of fixation, the first detectable photosynthetic products are amino acids. CO_2 is used to form pyruvate by means of pyruvate synthetase reaction. Reducing power is provided by the oxidation of H_2S. Reducing power obtained thus will reduce ferrosoxin (Fd). Acetyl CoA (of Krebs cycle), accepts CO_2 and is reduced by Ferrodoxin to yield pyruvate. The reactions take place as follows –

1. *Formation of pyruvate*

Acetyl CoA + CO_2 + Ferredoxin (reduced) —— pyruvate + CoA + Ferredoxin (oxidized)

2. *Formation of Oxaloacetate*

Pyruvate + ATP + CO_2 —— oxaloacetate + ADP + Pi

Oxaloacetate enters the reversed tricarboxylic acid cycle to yield succinyl CoA.

3. *Formation of* α *Ketoglutarate*

Succinyl CoA + CO_2 + Ferredoxin (reduced) —— α Ketoglutarate + CoA + Ferredoxin (Oxidized)

4. *Formation of Ctirate*

α Ketoglutarate is converted into citrate through oxalosuccinate. Citrate the, splits into oxaloacetic acid and acetate.

α Ketoglutarate CO_2 —— oxalosuccinate —— citrate —— oxaloacetic acid + Acetate

The above reactions are generally irreversible, but because of string reducing potential of the Ferrodoxin, the reaction can be reversed. As these bacteria can generate reduced ferrodoxin by means of photosynthesis, it is called reductive carboxylic acid cycle (Evans, Buchanan, Arnon, 1966).

The net result for every turn of the cycle is that four molecules of CO_2 are incorporated (reductive fixatin), and on molecule of oxaloacetate is synthesized. Besides ferredoxin, the cycle requires all other cofactors of Krebs cycle in reduced form. Three molecules of ATP ar required as follows for this fixation - 1. Activation of acetate 2. Carboxylation of pyruvate and 3. Activation of succinate.

Organic carbon assimilation pathway

Certain types of bacteria use organic compounds to provide CO_2 for fixation (not atmospheric CO_2) Stainer (1961), opined that a part of the organic carbon is directly assimilated. Accordingly to her, photoassimilated organic substrates (from organic polymers) such as acetate, is directly converted into poly β hydroxybutyrate as follows :

According to Gaffron (1963), however, during photoassimilation CO_2 reduction occurs only when excess reductants are available. Under anaerobic conditions, photometabolism of organic acids is strictly light dependent, and may be dependent on light generated ATP.

COMPARISON BETWEEN BACTERIA AND GREEN PLANTS (in terms of photosynthesis)

In the photosynthetic activity, there are a number of similarities and a few differences between green plants and photosynthetic bacteria. It is believed that photosynthesis in green plants is evolved from that found in simple green and purple bacteria. There is difference between green plants and bacteria in the localization of the photosynthetic pigments; while well organized chloroplasts are there in the former, in the latter, pigments are localized in membranous vesicles. (There is infact an opinion that the chloroplasts with their own DNA can be regarded as some primitive phototrophic organisms which have been trapped by cells of plants in the course of evolution).

The photosynthetic pigments viz., the chlorophylls of bacteria are different from those of higher plants in terms of their absorption spectra. It should be pointed out, however, that the basic chemical framework of all chlorophylls is the same. In carotenoid composition, however, green plants and bacteria differ– open aliphatic chain Carotenoids are seen in the latter, while the former have bicyclic Carotenoids.

In terms of the photochemical processes, bacteria are *anoxygenic,* while green plants (Cyanobacteria also) are *oxygenic.* This is due to the fact that in bacteria, the reducing power comes from H_2S and other inorganic and organic compounds, while in green plants water gets reduced and oxygen is released in the process. It is an enigma as to why bacteria have not chosen H_2O as a source of reducing power.

The source of cellular carbon is always atmospheric CO_2 in green plants, while it may be CO_2 or other organic compounds in bacteria.

There are a number of similarities also between bacteria and green plants in pigment composition. *Chl a* is present in both, though bacteriochlorophyll has two more hydrogen than green plant chlorophyll. Both chlorophylls are formed in membrane bound particles in the cell. In both, chlorophyll systems reaction centres exist with antenna pigments channelising light energy for harvesting.

In both green plants and bacteria, the overall photosynthesis is an oxidation reduction reaction. A reductant H_2A gets oxidized and provides the reducing power to convert carbon to cellular carbon. Oxidized 'A' is released in the process. Cyclic and non-cyclic photophosphorylation is found in both. Electron transport chain leading to the formation of ATP and reduced NAD compound is found in both green plants and bacteria. Carbon assimilation is similar as well as different in both the groups with Calvin cycle being common.

There is some difference in the photoreductant compound. It is $NADPH_2$ in green plants and $NADPH_2$ in photosynthetic bacteria. But the chromatophores of bacteria have an enzyme transhydrogenase and can photoreduce NADP in the presence of NAD. The following are the summarized similarities and differences between bacterial photosynthesis and green plant photosynthesis.

Similarities

1. Presence of chlorophyll *a*
2. Arrangement of photosynthetic pigments in membranous particles in the cell.
3. Presence of a definite reaction centre for harvesting light energy.
4. Accessory pigments channelise light energy towards reaction centre.
5. Occurrence of photophosphorylation, cyclic as well as non-cyclic.
6. Electron carriers of the chain are almost similar.
7. Light energy produces the reducing power.
8. Overall photosynthesis is an oxidation–reduction process.

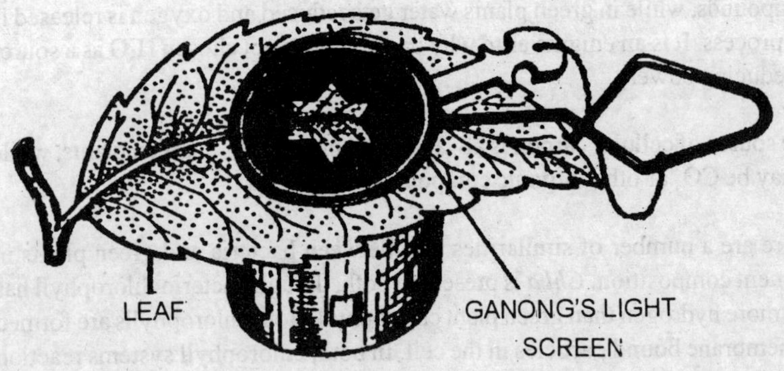

Fig. 5.30 Photosynthesis
Experiment to demonstrate the necessity of light for photosynthesis

Differences between photosynthesis found in bacteria and green plants

	Bacteria	Green Plants
1.	Bacteria have no distinct chloroplast	Chloroplast is well developed
2.	Bacteria absorb light of longer wavelength (800-900 nm or infra red)	These absorb light of relatively shorter wavelength (450-700 nm).
3.	P_{890} is the reaction centre	P_{680} and P_{700} are two reaction centres
4.	Chlorophyll *a* is absent and bacteriochlorophyll takes its functions	Chlorophyll *a* is present which converts radiant into chemical energy.
5.	The carotenoids are open chain alephatic type	Carotenoids are bicyclic
6.	Oxygen is not evolved during the product (anoxygenic photosynthesis)	Oxygen is evolved as by product(oxygenic photosynthesis).
7.	Water does not serve as a source of reducing power (electron donor)	Water serves as source of reducing power.

8.	Bacteria can use CO_2 as well as organic compounds as a source of carbon	Only CO_2 is the source of carbon
9.	The photoreductant is $NADH_2$	The photoreductant is $NADPH_2$.
10.	The process occurs in the presence of light but in the absence of O_2	Both are present during thc process
11.	Emerson effect is not found	Emerson effect is found dominant
12.	Cyclic photophosphorylation is dominant	Non-cyclic photophosphorylation isdominant
13.	Plastocyanin is absent	Plastocyanin is present
14.	Lycopene is found higher	Lycopene is not found in the chloroplast of plants.

6

RESPIRATION

The most basic requirement of all living organisms is *energy*. Without energy, no life activity is possible. The solar energy that is trapped by green plants are be stored in cells in the form of carbohydrates and other nutrients. These are oxidised and the energy so released utilized to carry on the vital processes of living beings.

The process of energy release by the (oxidation of) substrate material is called respiration. The term respiration, which has been in usage since a long time is derived from the latin word *respirare* (= breathe). Breathing as a phenomenon of life has been noticed in animals long before its relation to respiration and its universal occurrence in all forms of life.

Breathing and respiration

Breathing and respiration are often used as synonymous. But it is not correct. Breathing as a separate process is noticed only in few organisms (animals), while respiration occurs in all organisms. Further, breathing is a physical process, while respiration is a chemical process taking place in all the cells. Breathing may be regarded as the last step found only in aerobic respiration, where atmospheric oxygen participates. There can be respiration without breathing, but the reverse is not possible.

Respiration and combustion

Respiration is often compared to the process of combution, because in both the instances the substrate material gets burnt (oxidized) up. But the process of respiration differs from mere combustion in many respects. In combusion, the burning up or oxidation of the substrate is sudden, while in respiration it is stepwise. In combustion, a large amount of energy is wasted as heat, while in respiration it is channelised and stored up for further usage. Hence, respiration may be termed as an orderly combustion found in living organisms.

Cellular respiration

In a multicellular organism where there are thousands of cells, how does energy reach each cell? Each cell obtains its own energy by the oxidation of substrate material present within. Hence, respiration takes place in all the cells wherever they may be present in animals and plants. This is called cellular respiration.

Cellular respiration is a complex process which includes various aspects. These are the following.

(a) Absorption or oxygen (not in all)

(b) Degradation of substrates such as carbohydrates to CO_2 and H_2O (oxidation).

(c) Energy is released in the process of oxidation; some amount will also be lost in the form of heat.

(d) Formation of many intermediate products which play different roles in metabolism.

(e) There will be some loss of weight due to the process of oxidation. From the point of view of plant metabolism, cellular respiration is of paramount significance as it is the basic biological process that provides energy for all the vital activities. In short, the following aspects may be mentioned as indicating the significance of respiration.

   (i) Energy released in respiration is used for various metabolic processes. In fact, most of the physiological activities are driven by respiratory energy.

   (ii) Many intermediate compounds are formed which play a key role in the intermediate metabolism.

   (iii) Insoluble food is converted into soluble form.

   (iv) Potential energy is converted into kinetic energy.

   (v) CO_2, which is released during respiration is an important step in maintaining the carbon balance of nature.

The overall formula for respiration is as follows :

$$C_6H_{12}O_6 + H_2O + 6O_2\ 6CO_2 + 12\,H_2O + 686\text{ K cal energy.}$$

Respiratory substrate

The energy rich material (food substrate), that is used for the purpose of oxidation is called the *respiratory substrate*. In other words a respiratory substrate is defined as "any organic plant constituent oxidized partially or completely with the concomitant release of energy together with CO_2 and H_2O".

Plants use a variety of organic constitutents as their respiratory substrates. These may be a variety of carbohydrates (hexose sugars, disaccharides, polysaccharides), organic acids, fats etc. Under certain circumstances like starvation, even proteins may be used as respiratory substrates.

Among the carbohydrates, sucrose and starch are the principal respiratory substrates. When simple monosaccharides like glucose and fructose are available, they are directly utilized as respiratory substrates, while sucrose and starch are first hydrolysed to simple sugars before being channeled into respirtory metabolism.

In certain seeds like castor bean, fats are reserve food materials. They are first broken down into fatty acids and glycerol. As already mentioned, proteins will be channeled into respiratory metabolism under special circumstances such as starvation. Blackmann, who has studied respiration in detail, observed that in leaves kept in darkness, the respiratory rate which will be normal initially, declines, but does not stop altogether. Blackmann termed the normal respiration (when carbohydrates are utilized), as *floating respiration* and the slowed down rate of respiration as *protoplasmic respiration,* when the essential proteins are oxidised to keep the cells alive. In the long run, however this is harmful to the cells. When proteins are the respiratory substrates, they are first broken down into amino acids and these by deamination get converted into keto acids and enter the respiratory chain.

Respiratory Quotient (RQ)

RQ may be defined as *"the ratio between the volume of carbon dioxide given out and oxygen taken in simultaneously, by a given weight of the tissue in a given period of time, at standard temperature and pressure".*

RQ may be calculated as follows

$$RQ = \frac{\text{Volume of } CO_2 \text{ evolved}}{\text{Volume of } O_2 \text{ consumed}}$$

The value of RQ depends upon the nature of the respiratory substrate, amount of oxygen present (in the substrate), and the extent to which the substrate gets broken. Normally, the value of RQ should be unity (1), but many deviations are noticed due to the alterations in the levels of oxidation and reduction. The value of RQ also depends upon whether all the oxygen consumed is used up only for respiration or some of it is utilized for other purposes also. It is for these reasons that RQ may be unity, less than unity or even more than unity. RQ value also gives an idea of the respiratory substrate that is being used. The following are some RQ values when different respiratory substrates are involved.

RQ for Carbohydrates

Carbohydrates are the principal respiratory substances in a large majority of the organisms. These include a variety of monosaccharides (glucose, fructose etc), disaccharides (sucrose) and polysaccharides (starch, inulin etc). When simple sugars are available they are directly, channelised into the respiratory stream, but when poly and disaccharides are involved, they are first hydrolysed into monosaccharides.

When hexose sugars are the respiratory substrate, volume of CO_2 evolved equals volume of O_2 used, and hence the value will be unity. This will be illustrated by the following equation.

$$C_6H_{12}O_6 \rightarrow 6O_2 + H_2O$$

$$RQ = \frac{CO_2}{O_2} = \frac{6CO_2}{6O_2}$$

It has to be noted that not in all cases when sugar is the respiratory substrate, the RQ will be unity. There will be many deviations due to the following reasons.

(i) Incomplete oxidation of sugars (as in anaerobic respiration).

(ii) Involvement of many reductive events such as sulphate and nitrate reduction.

(iii) Oxidation and decarboxylation processes unrelated to respiration.

(iv) Retention of CO_2 within the cell fluids instead of releasing it to the exterior.

(v) Non photosynthetic carboxylation reactions resulting in accumulation of these products as seen in succulent plants.

(vi) Metabolic utilization of CO_2

RQ for Fats

Fats are not generally found in vegetative parts, but they are found as storage products in many of the angiosperm seeds like sunflower, castor, niger seed, groundnut etc. At the time of the germination of seeds, a major amount of fat is converted into carbohydrates, while the rest is used in respiration.

Fats are poor in oxygen content, and the ratio of oxygen to carbon is always less in fats when compared with carbohydrates. Fats are rich in hydrogen and as a result they require more O_2 for complete oxidation. In addition to this, fats are not directly oxidised. They are first broken into *fatty acids* and *glycerol,* which then act as respiratory substrates. Oxygen is required for the breakdown of fats into fatty acids and glycerol. As a result, for the overall respiratory process (when fats are involved), more oxygen is required. Hence, the value of RQ will be less than unity.

(a) $C_{18}H_{36}O_2 + 26O_2 \rightarrow 18\,CO_2 + 18H_2O$
(Stearic acid)

$$RQ = \frac{26\,O_2}{18CO_2} = 0.69$$

(b) $C_{57}H_{16}4O_6 + 80O_2 \rightarrow 57CO_2 + 52H_2O$
(Triolein fat)

$$RQ = \frac{57}{80} = 0.7$$

When simple fatty acids like acetic acids are involved, RQ will be unity as seen below.

(c) $CH_3COOHY + 2O_2 \rightarrow 2CO_2 + 2H_2O$
(Acetic acid)

$$RQ = \frac{20_2}{2\,CO_2} = 1.\text{ (unity)}$$

(d) $2C_{51}H_{98}O_6 + 145\,O_2 \rightarrow 102\,CO_2 + 98H_2O$
(Tripalmitin)

$$RQ = \frac{102}{145} = 0.7$$

Interestingly, fats liberate approximately three times more energy when compared with Carbohydrates. While a gram of carbohydrates yeilds about 3.8 K cal, the same amount of fat produces approximately 9.1 K cal of energy. This may be the reason as to why the seeds prefer to store reserve food material in the form of oil, more amount of energy can be obtained by less amount of food (it requires less space).

RQ for Proteins

Proteins and amino acids serve as respiratory substrates only under certain situations like starvation etc. Like fats, proteins also have less oxygen and the volume of oxygen to carbon is low. When proteins are hydrolysed, they require more oxygen for complete oxidation as a result of which RQ value will be less than unity. The RQ value for proteins and their derivatives varies between 0.5–0.8; very often it will be 0.79.

The value of RQ may be 0.99 when ammonia is released or it may be 0.79 when amides are formed.

RQ for Organic acids

The value of RQ is more than unity when organic acids are used as respiratory substrates. This is due to the fact that in organic acids, the proportion of oxygen in relation to carbon is always more, or less of O_2 is required for oxidation. The following are some of the examples.

(a) $C_4H6O_5 + 3O_2 \rightarrow 4CO2 + 3H2O$
(malic acid)

$$RQ = \frac{4CO_2}{3O_2} = 1.3$$

(b) $2(COOH)_2 + O2 \rightarrow 4CO_2 + 2H_2O$
(oxalic acid)

$$RQ = \frac{4}{1} = 4.0$$

(c) $2C_4H_6O_6 + 5O_2 \rightarrow 8CO_2 + H_2O$
(tartaric acid)

$$RQ = \frac{8}{5} = 1.6$$

(d) $2C6H8O7 + 9O2 \rightarrow 12CO2 + 8H2O$
(Citric acid)

$$RQ = \frac{12}{9} = 1.33$$

RQ for succulent plants

In succulent xerophytes such as *Opuntia,* and many other plants (*Bryophyllum, Sedum* etc), carbohydrates are incommpletely oxidized resulting in the formation of intermediate compounds, there being no evolution of CO_2. As only O_2 is taken in and there is no release of CO_2, RQ is zero.

In succulent plants, stomata are open during night, O_2 is taken in with the result carbohydrates are incompletely oxidized, while during day time, the organic acids are completely oxidized releasing CO_2. But this CO_2 is not released as it is used for photosynthesis. Hence, with no evolution of CO_2 RQ will be zero.

$$2C_6H_{12}O_6 + 3O_2 \rightarrow 3C_4H_6O_5 + 3H_2O$$

$$RQ = \frac{CO_2}{O_2} = \frac{O}{3} = 0.0$$

RQ when O_2 is used for other activities

O_2 is also utilized for various other processes like converon of fats to carbohydrates, synthesis of anthocyanins etc. In such cases, the amount of CO_2 evolved will not correspond to the amount of O_2 consumed. Here RQ falls below unity.

RQ for anaerobic respiration

Certain tissues respire anaerobically (no participation of atmospheric, O_2), in which case only CO_2 is evolved without the absorption of O_2. In such circumstances, RQ will be more than unity.

$$C6H12O6 \rightarrow 2C2H5OH2 + 2CO2$$

$$RQ = \frac{CO2}{O_2} = \frac{2}{O} = 2$$

The following is a list of RQ values for some of the substrates occuring in various plant parts.

No.	Plant part	Substrate	SQ value
1.	Leaves	Carbohydrates	1.0
2.	*Opunitia* shoots	Organic acids	0.03
3.	Seeds	Starch	1.0
4.	Linsed	Fat	0.64
5.	Buckwheat seed	Protein	0.5
6.	Pea Seeds	Carbohydrates	1.5-2.4

Measurement of RQ

RQ can be measured with the help of *Ganong's respiroscope.* The apparatus essentially consists of a graduated tube with a bulb at the upper end and a levelling tube. The two tubes are connected at their base with a long rubber tube which assumes a 'U' shape. The bulb in the graduated tube has a glass stopper fitted into its neck. Both the neck and the stopper have holes which can be brought into a straight line by twisting the stopper. The apparatus is fitted into an iron or wooden stand, and filled with mercury (instead of Hg, saline solution also can be used). Respirable material such as plant parts or germinating seeds are kept in the bulb and the stopper is rotated to allow the air enter into the bulb. Mercury level in both the tubes is brought at the same level. Now the stopper is rotated to cut off the communication.

The initial level of Hg is noted. If the respiratory substrate is carbohydrate, CO_2 released will be equal to O_2 observed, so that there is neither rise nor decrease in the level of Hg. If the substrate is fat or protein, CO_2 released will be less than O_2 consumed and therefore a vaccum will be created in the chamber. This results in the increase of the level of Hg.

If the CO_2 released is more than O_2 used up, the initial level of Hg will fall after the experiment. RQ in this case will be more than one. RQ may be calculated with the help of the following formulae.

$$RQ = \frac{v^2}{v^1 + v^2}$$

v^1 = Initial rise of Hg; v^2 = further rise in Hg level after introduction of caustic potash. This is understood to find out the excess release of CO_2 if any.

$$RQ = \frac{v^2}{v^2 --- v^2}$$

v^1 = Fall in the initial level of Hg. v^2 = Rise in level after introduction of caustic potash.

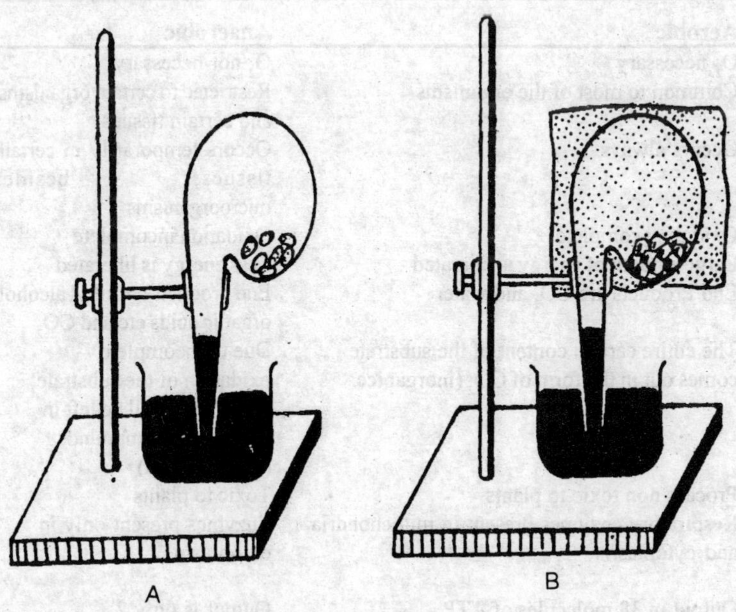

Fig. 6.1 Respiration

Demonstration of value of RQ for carbohydrates. **A.** Initial leel, **B.** Ater 24 hours

TYPES OF RESPIRATION

The involvement of atmospheric oxygen during respiration constitutes the basis for its categorization into aerobic and anaerobic types.

Aerobic respiration

The respiratory substrate is completely broken down into CO_2 and H_2O with the participation of atmospheric O_2.

$$C_6H_{12}O_6 + 6O_2 \rightarrow 6CO_2 + 6H_2O + 686K\ Cal.$$

Anaerobic respiration

Here there is no participation of atmospheric oxygen. Food substances are incompletely oxidized. Hence, all the carbon in the substrate will not be released in the form of CO_2. The end products are not only CO_2, but intermediate products like ethyl alcohol, acids etc.

$$C_6H_{12}O \rightarrow 2C_2H_5OH + 2CO_2 + 56K\ Cal.$$

Besides the above, there are some more differences between aerobic and anaerobic respiration. They are listed below.

	Aerobic	Anaerobic
1.	O_2 necessary	O_2 not necessary
2.	Common to most of the organisms	Restricted to certain organisms, and certain tissues.
3.	Occurs always	Occurs temporarily in certain tissues, besides microorganisms
4.	Oxidation complete	Oxidation incomplete
5.	Large amount of energy is liberated	Less energy is liberated
6.	End Products are CO_2 and water	End products are ethyl alcohol, organic acids etc and CO_2
7.	The entire carbon content of the substrate comes out in the form of CO_2 (inorganicc	Due to incomplete oxidation of the substrate, carbon will still be left in the organic compounds (ethyl alcohol)
8.	Process non toxic to plants	Toxic to plants
9.	Respiratory enzymes present in mitochondria and cytoplasm.	Enzymes present only in cytoplasm
10.	Output in 38 molecules of ATP	Output is only 2 molecules of ATP

MECHANISM OF RESPIRATION

As has already been pointed out, during respiration the substrate material is broken down with release of energy, CO_2 and H_2O.

In the mechanism, we will be studying the various stages of the breakdown of the substrate material, enzymes involved etc., and account for the release of CO_2 and energy released in the form of molecules of ATP. In the foregoing paragraphs, various steps involved in the breakdown of carbohydrates are given.

Broadly categorizing, the following steps are identifiable in the oxidative breakdown of carbohydrates.

1. Hydrolysis 2. Glycolysis 3. Pyruvic acid oxidation 4. Krebs cycle or citric acid cycle and 5. Electron transport system or oxidative phosphorylation.
The mechanism of anaerobic of anaerobic respiration consists of pyruvic acid.

1. Glycolysis and 2. Anaerobic oxidation of pyruvic acid.
From the above amount it can be understood that the step glycolysis is common for both aerobic and anaerobic respiration.

HYDROLYSIS

When starch is the reserve food material, it should first be hydrolysed into simple sugars before it can enter into the respiratory cycle. The process of breakdown of starch into simple sugars with the addition of water is called hydrolysis.

GLYCOLYSIS

During glycolysis, one molecule of glucose will be broken into two molecules of pyruvic acid. Originally, the name glycolysis was meant for the breakdown of glycogen (*lysis* or breakdown of glycogen) found in muscle cells. Enzymes responsible for glycolysis are present in the cytoplasm. As will be seen later, no participation of atmospheric O_2 is required for the formation of pyruvic acid. Glycolysis is the common pathway for both aerobic and anaerobic respiration.

Glycolytic breakdown of glucose into pyruvic acid is also known as the **Embden - Meyerhof - Parnas** pathway (EMP pathway), after three scientists who discovered the various steps of glycolysis. EMP pathway is also called cytoplasmic respiration as the whole process of glycolysis takes place only in cytosol (cytoplasm minus organeloles). EMP pathway was first worked out in muscle cells and yeast, and later it was observed that it is common for all the organisms. The overall reaction of glycolysis may be summed up as follows.

$C_6H_{12}O_6 \rightarrow 2CH_3COCOOH + 4H$

The details of the above reaction are given below.

Phosphorylation

Any sugar that has to enter into the glycolysis pathway must first undergo phosphorylation (i.e., a phosphate group has to be added). In this process, ATP participates and donates a phosphate group (there will be utilization of a molecule of ATP).

(i) Glucose + ATP Glucose 6 Phosphate + ADP

The reaction is catalyzed by the enzyme *hexokinase* with Mg ++ as the cofactor. This reaction was first worked out by Meyerhof in 1927.

Glucose 6 phosphate may be produced by another means also, if starch is the original food material. It is as follows.

Starch + H3PO4 → Glucose 1 phosphate

The enzyme phosphorylase catalyses this reaction. In the next step

Glucose 1 phosphate → Glucose 6 phosphate

In the above reaction Glucose 1 phosphate changes into Glucose 6 phosphagte i.e., the phosphate group attached to the first carbon atom is shifted to the sixth carbon atom. The reaction is catalyzed by the enzyme *Phosphoglucomutase.*

Isomerization

In the step, glucose 6 phosphate isomerises into Fructose 6 phosphate under the influence of the enzyme *Phosphohexoisomerase.*

Glucose 6 phosphate → Fructose 6 phosphate

Fructose 6 phosphate may be produced from fructose directly when the latter is available.

$$\text{Fructose + ATP} \xrightarrow[\text{Mg ++}]{\textit{Hexokinase}} \text{Fructose 6 phosphate}$$

Formation of Fructose 1.6. diphosphate

This is a key reaction in the initial breakdown of glucose. Fructose 6 phosphate utilizes a molecule of ATP to get converted to Fructose 1.6 diphosphate under the influence of the enzyme *Phosphohexokinase.*

Fructose 6 phosphate + ATP → Fructose 1.6 Diphosphate + ADP

The reactions of phosphorylation leading to the formation of Fructose 1.6 diphosphate may be summarized as follows.

$$\text{(i) Glucose + ATP} \xrightarrow{\textit{Hexokinase}} \text{Glucose 6. Phosphate + ADP}$$

Mg ++

(ii) Glucose 6 Phosphate —*Phosphohexoisomerase*→ Fructose 6 Phosphate

(iii) Fructose 6 Phosphate + ATP —*Phosphohexokinase*, Mg ++→ Fructose 1.6 Diphosphate + ADP

Cleavage of Fructose 1.6 dip

Further reactions in glycolysis begin with the breaking up of Fructose 1.6 diphosphate into two molecules of triose under the influence of the enzyme *Aldolase.* The presence of this enzyme in plant tissues was first discovered by Tewfik and Stumf (1949).

1. Fructose 1.6 diphosphate ______ Glyceraldehyde 3 phosphate + Dihydroxyacetone Phosphate.

97% of Fructose 1.6 diphosphate gets converted into glyceraldehyde 3 phosphate, while about 3% will change into dihydroxyacetone phosphate. Dihydroxyacetone phosphate isomerises into glyceraldehyde 3 phosphate under the influence of the enzyme *phosphotriose isomerase.* The reaction may be represented as follows.

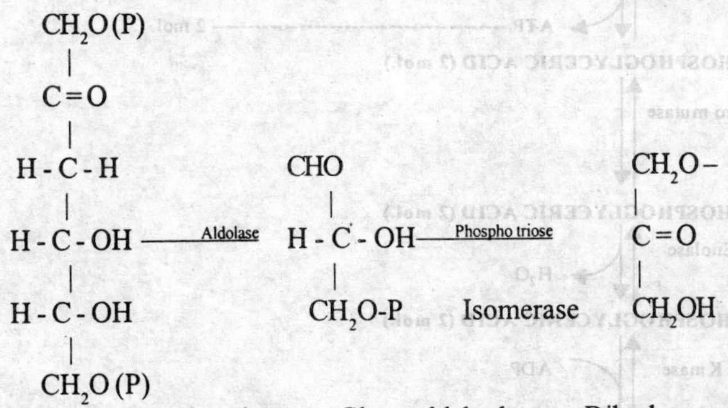

Fructose 1.6 Diphosphate Glyceraldehyde 3 - Phosphate Dihydroxyacetone Phosphate

To sum up, Fructose 1.6 diphosphate (6C) breaks up into 2 molecules of glyceraldehyde 3 phosphate (3C). From now on the reactions are explained for a molecule of triose (glyceraldehyde 3 phosphate). While accounting for hexose, each reaction has to be doubled. In the next step, glyceraldehyde 3 phosphate

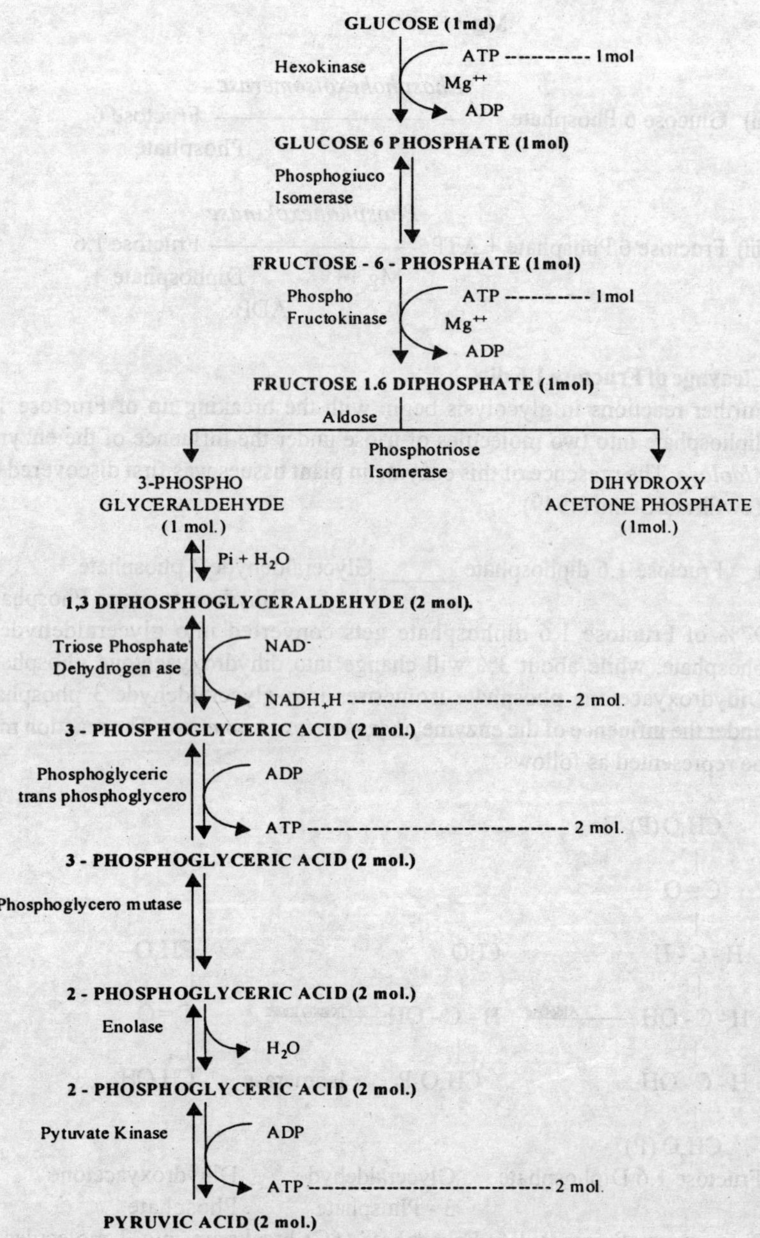

Fig. 6.2 Respiration

Glycolysis - Various reactions leading to the formation of Pyruvic acid

is oxidized (hydrogen removed) with the help of the enzyme *phosphoglyceraldenhyde dehydrogenase.* Inorganic phosphate and NAD participate in this reaction. 2 electrons and two protons (H+) are released from glyceraldehyde 3. Phosphate and the energy is utilized to link the inorganic phosphate to the oxidized glyceraldehyde 3 phosphate to form 1,3 - diphosphoglyceric acid.

Glyceradehyde 3 phosphate + H_3PO_4 + NAD
———————————— 1.3 Diphosphoglyceric acid +
NADH + H +

As may be noticed, while the electrons provide the energy necessary for the linking of inorganic phosphate, the H + goes for the reduction of NAD to NADH +H^+

```
CHO                                   O
|                                     ||
H - C - OH + H2PO4 + NAD              C - O - P
|                                     |
CH2OP                                 H - C - OH + NADH + H+
                                      |
                                      CH2O P
lyceraldehyde - 3 Phosphate           1.3 - Diphosphoglyceric acid
```

3. 1,3 - Diphosphoglyceric acid is transformed into 3 phospoglyceric acid when one of the phosphate groups is released, which is accepted by ADP to synthesize a molecule of ATP. *Phosphoglyceric kinase* enzyme catalyzes this reaction.

The enzyme has Mg^{++} as the cofactor

1.3 Diphosphoglyceric acid + ADP 3 Phosphoglyceric acid + ATP

```
        O
        | |
        C - O - P                             COOH
        |            Phosphogyceric           |
H - C -- OH + ADP    ——————————————    H - C - OH + ATP
     CH2O - P        Kinase + Mg++           CH2O - P
```

4. 3 Phosphoglyceric acid gets transformed into 2 phosphologlyceric acid (phosphate group is shifted from the 3rd carbon position to the 2nd carbon position), under the influence of the enzyme *Phosphoglyceromutase.*

```
COOH                                          COOH
|          Phosphoglyceromutase               |
```

$$\begin{array}{c} H-C-OH \\ | \\ CH_2O-P \end{array} \xrightarrow{\text{-mutase}} \begin{array}{c} H-C-O-P \\ | \\ CH_2OH \end{array}$$

3. Phosphoglyceric acid

5. 2 - Phosphoglyceric acid (2PGA) gets converted into phosphoenolpyruvic acid (PEP) under the influence of the enzyme, *enolase* with Mg^{++} ions as cofactor. A molecule of water is eliminated from 2PGA in the process with the result an energy rich (-P) centre is created.

$$\begin{array}{c} COOH \\ | \\ H-C-O-P \\ | \\ CH_2OH \\ (2\,PGA) \end{array} \xrightarrow[Mg^{++}]{\textit{Enolase}} \begin{array}{c} COOH \\ | \\ H-C-O-P+H_2O \\ | \\ CH_2 \\ (2\,PEP) \end{array}$$

6. In the molecule of PEP, much of the energy is centered round the phosphate group. This energy rich phosphate group is removed from PEP, which is accepted by ADP to produce a molecule of ATP. In the process, the dephosphorylated PEP becomes converted to pyruvic acid. This reaction is catalyzed by the enzyme *pyruvate kinase* with Mg++ and K+ ions as cofactors.

$$\begin{array}{c} COOH \\ | \\ H-C-O-P+ADP \\ | \\ CH_2 \end{array} \xrightarrow[Mg^{++}K^{++}]{\textit{pyruvate kinase}} \begin{array}{c} COOH \\ | \\ C=O+ATP \\ | \\ CH_3 \\ \text{(Pyruvic acid)} \end{array}$$

With the formation of pyruvic acid, the reactions of glycolysis come to an end. One molecule of pyruvic acid is produced for every triose, hence for hexose, it should be two molecules of pyruvate.

ATP account during glycolysis

The production of pyruvic acid terminates glycolysis. When we briefly glance through the glycolytic reactions we realize that there are two important stages – (i) phosphorylation and (ii) oxidation of Fructose 1.6 diphosphate to pyruvic acid.

During phosphorylation of glucose to form Fructose 1.6 diphosphate, two molecules of ATP are used up. Thus during phosphorylation of glucose, 2 molecules of ATP are used.

In the second stage, at two stages ATP molecules are synthesized–once during the conversion of 1,3 - diphosphoglyceric acid to 3 phosphoglyceric acid (2 molecules - i.3., one for every triose and two for every hexose), and again during the conversion of phosphoenol pyruvic acid to pyruvic acid (2 molecules of ATP, reactions are to be doubled while accounting for hexose). Thus in all, towards the end of glycolysis 4 molecules of ATP are produced and two molecules of ATP are consumed (during phosphorlation). Hence, the net gain of ATP during glycolysis is only two molecules.

But the statement that the net gain of ATP during glycolysis is only two molecules is not completely correct, because during the oxidation of glyceraldehyde 3 phosphate to phosphoglyceric acid, 2 molecules of NADH + H + are produced. These two molecules pass through the electron transport chain producing a total of 6 (3 + 3) ATP molecules (Details explained later). Hence, the net gain of ATP during aerobic glycolysis is, 6 + 2 = 8 molecules of ATP.

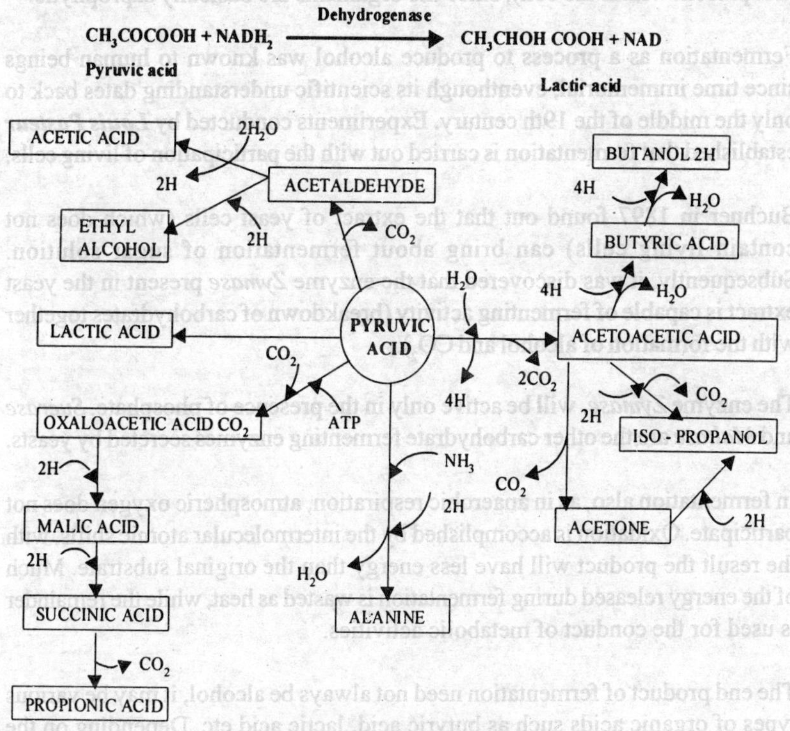

Fig. 6.3 Respiration
Anaerobic degradation of Pyruvic acid

Breakdown of pyruvic acid

Further breakdown of pyruvic acid depends upon the nature of the organism, type of respiratory pathway and the presence or absence of O_2. Broadly these breakdown are –

(i) Alcoholic fermentation
(ii) Lactic acid fermentation
(iii) Amino acid synthesis and
(iv) Aerobic breakdown to produce CO_2 and H_2O

Options (i) and (ii) can be carried out in the abscence of O_2 and (iv) is possible in the presence of O_2, which happens to be the major pathway. Option (iii) leads to the biosynthesis of alanine by transaminase reaction with glutamate. This is not discussed here as it is not relevant to the respiratory pathway.

FERMENTATION

Fermentation is essentially anaerobic respiration found in microorganisms like fungi and bacteria. In fermentation, the substrate for oxidation is extracellular (not present within the cell), since the organisms are basically saprophytic.

Fermentation as a process to produce alcohol was known to human beings since time immemorial, eventhough its scientific understanding dates back to only the middle of the 19th century. Experiments conducted by ***Louis Pasteur*** established that fermentation is carried out with the participation of living cells.

Buchner in 1897 found out that the extract of yeast cells (which does not contain living cells) can bring about fermentation of sugar solution. Subsequently, it was discovered that the enzyme *Zymase* present in the yeast extract is capable of fermenting activity (breakdown of carbohydrates together with the formation of alcohol and CO_2).

The enzyme *Zymase,* will be active only in the presence of phosphate. *Sucrase* and *Maltase* are the other carbohydrate fermenting enzymes secreted by yeasts.

In fermentation also, as in anaerobic respiration, atmospheric oxygen does not participate. Oxidation is accomplished by the intermolecular atomic shifts, with the result the product will have less energy than the original substrate. Much of the energy released during fermentation is wasted as heat, while the remainder is used for the conduct of metabolic activities.

The end product of fermentation need not always be alcohol, it may be various types of organic acids such as butyric acid, lactic acid etc. Depending on the end product, fermentation may be termed as alcoholic fermentation, lactic acid fermentation etc. The following are formulae for these fermentations.

(i) $C_6H_{12}O_6 \rightarrow 2C_2H_5OH + 2CO_2$ (ethyl alcohol)
(ii) $C_6H_{12}O_6 \rightarrow C_4H_8O_2 + 2H_2 + 2CO_2$ (butyric acid)
(iii) $C_6H_{12}O_6 \rightarrow 2_3H_6O_3$ (Lactic acid)
(iv) $C_2H_5OH + O_2 \rightarrow CH_3\ COOH + H_2O$ + energy (acetic acid)

In the formula (iv) given above, for the formation of acetic acid from ethyl alcohol, O_2 is necessary. Hence acetic acid fermentation is different from other types of fermentation.

Fermentation is carried out by microorganisms, a point which has been emphasized earlier. The following is a list of organism which carry out the various types of acid fermentation.

	Organism	Type of fermentation (based on end product)
1.	*Bacillus butyricus and Clostridium butyricum*	Butyric acid
2.	*Bactyerium lactic acid*	Lactic acid
3.	*Acetobacter aceti*	Acetic acid

The brief description of fermentation given above may now be followed by certain details of a few specific types of fermentation and a mechanism of regeneration of ATP.

ATP Generation by fermentation

During the fermentation of the glucose by microorganisms about 1–4 molecules of ATP may be produced. While there are large numbers of end products of fermentation starting from glucose, it is surprising that there are only a few reactions concerned with energy conversions of substrate level phosphorylation. Of these, the three most important reactions are discussed below.

1. 1m3 - Bisphosphoglycerate + ADP ———— 3 - Phosphoglycerate + ATP
2. Phosphoglycerate + ADP ———— Pyruvate + ATP
3. Acetyphosphateor Butyrylphosphate + ADP ———— Acetate or Butyrate + ATP

Most of the fermenting microorganisms use only reaction 1 and 2 above, and use pyruvate compounds synthesised from acetyl - Co - A as the necessary hydrogen acceptors. In the above, reaction 1 is catalysed by phosphoglycerate kinase and reaction 2 is catalysed by pyruvate kinase, during the fermentation of one molecule of glycose only two or a maximum of 4 molecules of ATP are generated and the following products may be formed: lactate, ehtanol, acetone, butyrate, n-butanol, 2-propoanol, 2,3 - butanediol, caproate, acetate, carbon dioxide, and molecular hydrogen.

In reaction 3 given, above the enzyme involved is acetate kinase, and it offers the advantage of additional ATP regeneration. Acteyl phosphate is produced from acetyl - Co - A by phosphotransacetylase.

acetyl - Co - A + P1 ——— acetylphosphate + Co - A

Acetyl phosphate may also be produced from sugar phosphates such as fructose - 6 - phosphate via reactions catalysed by phospoketolase.

The ability of microbes to utilise acetate kinase depends on whether they can evolve molecular hydrogen. No hydrogen acceptors need to be synthesised for reducing equavalents that may be liberated as molecular hydrogen. Fermentations may be basically classified into two types :

1. Alcoholic fermentation,
2. Acid fementation

1. Alcoholic fermentation

When respiratory substrates are sugars, during anaerobic respiration many of the microbes ferment them and produce ethanol (ethyl alcohol). The wide variety of microorganisms are involved in ethanolic fermentation. The chief ethanolic producers are yeasts (particularly strains of *Saccharomyces cerevisiae)* and fungi. Yeasts also respire aerobically but in the absence of air, they ferment carbohydrates into ethanol and carbon dioxide. Among bacteria also a number of anaerobes or facultative aerobes produce ethanol as the main product during fermentation of sugars. Unusuall accumulation of ethanol can occur even in plants under anaerobic conditions.

Ethanol formation of yeasts

The fermentation process was infact studied in great detail in yeasts. The conversion of glucose to ethanol was formulated in 1815 by Guy – Lussac as follows.

$$C_6H_{12}O_6 \text{ ——— } 2CO_2 + 2C_2H_5OH$$

The normal fermentation of glucose by yeasts

The breakdown of glucose to ethanol and carbon dioxide by yeasts proceeds via the usual glycolytic pathway. From their onwards, the conversion of pyruvate to ethanol comprises few steps. In the first step, pyruvate is decarboxylated by pyruvate decarboxylase with the participation of thiamine pyrophosphate. As a result of this, a molecule of acetaldehyde is formed which is then reduced to ethanol by alcohol dehydrogenase with $NADH_2$ as the hydrogen donor. In this hydrogen transfer, the hydrogen obtained during the dehydrogenation of triphosphate is utilised. Thus the oxidation reduction balance in maintained.

History of yeast fermentation

Louis Pasteur was perhaps the first to have carried out experiments on fermentation by microorganims. It was Buchmer and Hahn (1896), however, who connected the fermenting activity with yeast. They observed that the extract obtained from Brewer's yeast by grinding with kieselguhr (diatomaceous earth), and sand produced gas bubbles after the addition of sugar. This was perhaps a first example of biochemical reaction *in vitro.* Harden and Young (1906), made the discovery that inorganic phosphate is a requirement for the fermentation of glucose by yeast extract and it is incorporated in fructose - 1,6 - bisphosphate. According to Harden and Young equation, fermentation of glucose by yeast is as follows.

$$2C_6H_{12}O_6 + 2Pi \longrightarrow 2CO_2 + 2C_2H_5OH + H_2O$$

+ fructose - 1,6 - bisphosphate

Fermentation formulae by Neuberg

The studies of Neuberg on fermentation are of great significance. He showed that yeast can ferment both glucose as well as pyruvate. The formation of intermediate compounds such as acetyldehyde could by demonstrated by trapping it with hydrogen sulphite. On addition of hydrogen sulphite to glucose fermenting yeast, acetyldehyde is precipitated as the addition product.

$$CH_3CHO + NaHSO_4 \longrightarrow CH_2 - CHOH - SO_3Na$$

Under the above conditions, glycerol is formed as a new product while the yield of ethanol and carbon dioxide gets reduced. This kind of fermentation has been exploited industrially for the production of glycerol. By trapping the acetyldehyde (using hydrogen sulphite), it is not allowed to function as hydrogen acceptor. As a result, dihydroxycetone phosphate becomes the hydrogen acceptor and is reduced to glycerol - 3 - phosphate which is then dephospharylated glucose + hydrogen sulphite $\longrightarrow$ glycerol + acetaldehyde - sulphite + CO_2

The above fermentation which is a modification of the normal ethanolic fermentation by yeasts is known as Neuberg's second fermentation. The principle of trapping a metabolite along the pathway has now become known as the trapping method and has been extensively used in biochemical research.

Addition of alkali to fermenting yeast extract, also leads to the formation of glycerol because acetyldehyde undergoes dismutation to ethanol and acetate, and hence cannot function as a hydrogen acceptor. This reaction is known as Neuberg third fermentation and is as follows:

$$2\ GLUCOSE + H_2O \text{ —— } ethanol + acetate + 2\ glycerol + 2CO_2$$

The normal ethanolic fermentation is referred to as Neuberg's first fermentation and the reaction is as follows.

$$C_6H_{12}O_6 \text{ ———— } 2C_2H_5OH + 2CO_2$$

Relationship of yeast to oxygen

Under anaerobic conditions, yeasts ferment very actively but they can hardly grow. This is because yeasts are aerobes. On admission of air, fermentation decreases in favour of aerobic respiration. As shown in Pasteur effect, fermentation can be completely supressed by vigrous aeration. Pasteur discovered this effect, more than 100 years ago during his investigations on wine making. Pasteur effect, however is not restricted only to yeast but has been found in all anaerobic cells. Aeration of yeasts supports growth, but leads to a decline in glucose consumption and in the reduction of ethanol and carbon dioxide. This observation appears to be sensible from the stand point of energy conversion and suggests an ingenious regulatory mechanism. One molecule of glucose anaerobically yields only two molecules of ATP in contrast to the aerobic yield of 38 molecules of ATP per glucose molecule. Thus, the cell adopts by regulation of substrate consumption to the energy gain obtainable under two conditions.

The Pasteur effect involves several regulatory mechanisms working tegether. One of these perhaps applies at the level of phosphorylation and can be explained by competition for ADP and P.

A second regulatory mechanism perhaps is also responsible for the Pasteur effect. The enzyme phosphofructokinase is allosterically inhibited by ATP.

Harden and Young's explanation of fermentation balance

According to Harden and Young, fructose, 1,6 biphosphate accumulates due to lack to ATP utilisation for energy requiring reactions in the cell - free system. This leads to an excess of ATP balance. The yeast extract does not have any phosphates activity so that ADP must be constantly regenerated by phospharylation of excess glucose or fructose - 6 phosphate.

Industrial uses of yeast

For industrial purposes, different strains of yeast are selected. These are of two types.

Bottom yeast and *Top yeast.* Bottom yeasts are weekly respiring and actively, fermenting ones, and hence are used in the brewing of beer. Top yeasts on the other hand are used for production of ethanol and wine making.

Baker's yeast (*Saccharomyces cerevisiae),* are mostly top yeasts and are mainly used in baker's to produce carbon dioxide to raise the dough, i.e, these are grown in tanks with good aeration. In this reaction, ethanol is produced as a byproduct.

Brewer's yeasts are mostly bottom yeasts and are mainly used in brewing of beer from barley.

Industrial alcohol is produced from the residue of cane sugar refining called molasses. Occassionally, potatoes also may be used as rawmaterial for fermentation. Alcohol may also be obtained by hydrolysed wood chips. Microbial species used for these fermentations are species of *Candida* and *Endomyces.*

Ethanol formation by bacteria

Not many bacteria have been studied and exploited for ethanol production as has been done in the case of yeast. Among all the bacteria examined so far, only in the case of *Sacrcina ventriculi,* the pathway of ethanol formation is similar to that of yeast. Another bacterium that seems to produce ethanol has been isolated from the juice of the Agave plant. This bacterium (*Zymomonas mobilist*) breaksdown glucose via 2-keto-3-deoxy-6-phosphogluconate pathway and splits the pyruvate to acetyldehyde and carbondioxide under the influence of the enzyme pyruvate decarboxylase. The acetyldehyde is further reduced to ethanol along with carbondioxide and some quantity of lactic acid. It has further been discovered that the agave alcohol contains carbon atoms 2,3,5 and 6 of glucose, whereas a yeast ethanol contains carbon atoms 1,2,5 and 6 of glucose.

Some of the bacteria belonging to Enterobacteriaceae and Clostridia are also known to produce ethanol as a product of secondary fermentation. In these microorganisms, however, the precurosor of ethanol namely acetyldehyde is not produced from pyruvate by pyruvate decarboxylase but originates by the reduction of acetal Co-A.

Some of the acid fermenting bacteria like *Leuconostoc mesenteriodes,* produce alcohol by an entirely different pathway. Glucose is first broken down by the early steps in the pentose phosphates cycle to produce pentose phosphate. This pentose phosphate (xylulose -5 - phosphate) is then acted upon by the

enzyme phosphoketolase to produce acteyl phosphate. This is reduced to ethanol by the enzymes acetaldehyde dehydrogenase and alcoholdehydrogenase. The remaining product of breakdown glyceroldehyde - 3 - phosphate is converted into pyruvate xylulose - 5 - phosphate + Pi — acetylphosphate
+ glyceraldehyde -3 - phosphate.

Acid fermentation

Anaerobic breakdown of sugars by certain microorganisms results in the formation of various types of acids such as lactic acid, acetic acid, butyric acid, propionic acid, formic acid etc. Some of these fermentation are discussed below in some detail.

Lactic acid fermentation

The lactic acid bacteria belong to the family *lactobacteriaceae* and includes a wide variety of rods and cocci. All the members of lactobacteriaceae whether it be the physical form have a characteristic physiology, are gram positive and non-sporing (with few exceptions). They are usually non-motile and have carbohydrate as the main energy source and produce lactic acid. Another family of bacteria which produce lactic acid is Enterobacteriaceae. The lactobacteriaceae, however, are obligate fermenters. Basically, lactobacteria are anaerobic but aerotolerant.

Generally, the members of lactobacteriaceae require many accessory factors in the culture medium in addition to the basic nutrients. Addition of vitamins, amino acids, purines and pyrimidines is necessary to cultivate lactobacteria. The culture medium is usually complex one containing yeast extract, tomato extract or even blood.

Generally, the metabolic machinery of lactobacteria appears to be quite defective as it cannot produce a number of compounds which are required for its growth and consequently they have to be added to the medium. Many microbiologists regard the lactobacteriaceae as "metabolic cripples", as they have lost the ability to synthesise a number of metabolants. This may be due to the fact that in the course of evolution they seem to have acquired an affinity to milk and other nutrient rich media. This has lead to a special metabolic mechanism in them to utilise lactose, an ability that most microorganisms do not possess. The ability to use lactose is also possessed by many intestinal bacteria of the coliform type.

Milk as such is produced as a nutrient factor by mammals and consists of lactose. It is not found in the plant kingdom. The ability of some microorganisms to utilise lactose may be regarded as an adaption to the environmental conditions prevailing in the mammalian digestive tract.

Lactose is a disaccharide which must be hydrolysed before it can enter the catabolic pathway to produce Hexoses. This reaction is catalysed by the enzyme β-galactosidase. This enzyme however, occurs only in, few bacteria. The product of hydrolysis mainly ngalactose is converted to glucose - 1 - phosphate after phosphorylation.

Occurrence of lactobacteria

Lactic acid bacteria are largely distributed in places where there is a generous amount of lactose, as such, they are hardly found in soil or water. The natural habitats of natural bacteria (Schlegel, 1993) are –

1. Milk and the places where milk is produced and processed (*Lactobacillus lactis, L. bulgaricus, L.Helveticus, L.casei, L.fermenttium, L.brevis, Lactococcus lactis, L.diacetilactis*).

2. Intact and rotting plants (*Lactobacillus plantarum, L.delbruackii, L.fermentum, L.brevis, Lactococcus lactis, Leuoconostoc mesenteriodes*)

3. Intestinal tracts and mucous membranes of animals and humans (*Lactobacillus acidophilus, Bifidobacterium, Enterococcus facealis, Streptococcus salivarius, S.bovis, S.oyogenes, S.pneumoniae*).

Classification of lactobacteria

They are classified into two basic categories based on their ability to ferment glucose into solely lactate or to other additional products. If the fermentation produce is only lactat the bacteria are said to be **homofermentative** and if the fermentation product is many other compounds in addition to lactate they are said to be **heterofermentative.** The following table gives a list of heterofermenting and homofermenting bacteria.

Cocci	**Rods**
Homodermentative : $C_6H_{12}O_6$ ——	$2CH_3$ - CHOH - COOH
Streptococci	Lactobacilli
Lactococcus lactis subsp. Lactis	Thermobacteria (temp. Opt 40^{0}C,do not grow at 15^{0}C)
Lactococcus lactis subsp. Lactis	subsp, Delbrueckii
var.diacetilactis	Lactobacillus Delbrueckii
subsp.cremoris	Subsp. Lactis
Enterococcus faecalis	Lactobacillus delbrueckii

Streptococcus salivarius
subsp.salivarius
Streptococcus salivarius
subsp.thermophilus
Streptococcus pyogenes

subsp.bulgaricus
Lactobacillus helveticus
Lactobacillus acidophilus
Lactobacillus salivarius
Streptobacteria (temp.opt. 30-37^{0}C, alwaysgrow at 15^{0}C)
Lactobacillus casei
Lactobacillus alimentaris
Lactobacillus coryniformis
Lactobacillus plantarum

Heterofermentative : $C_6H_{12}O_6 — CH_3$ - CHOH - COOH + CH_3 - $CH_2OH + CO_2$ (or CH3 - COOH)

Streptococci
Leuconostoc mesenteriodes
subsp. mesenteriodes
Leuconostoc mesenteriodes
subsp.dextranicum
Leuconostoc mesenteriodes
subsp.cremoris
Leuconosoc lactis

Lactobacilli
Lactobacillus bifermentans
Lactobacillus brevis
Lactobacillus fermentum
Lactobacillus kandleri
Lactobacillus viredescens

Homolactic fermentation

Homofermentative lactobacteria produce as the end product of metabolism almost pure lactate. They breakdown glucose via the fructose biphosphate pathway as they posses all the required enzymes. Aldolase, which is a key enzyme helps in the anaerobic breakdown of glucose to lactate. From glucose pyruvate is produced first, which is then reduced to lactate. The hydrogen necessary for reduction of pyruvate comes from the dehydrogenation of glyceraldehyde - 3 - phosphate. The type of lactate formed whether it is D, L or DL depends on the stereospecificity of the enzyme lactate dehydrogenase. Only the small fraction of the pyruvate is decarboxylated and converted to acetate, ethanol, carbon dioxide or possibly acetation. The extent of the formation of subsidiary products besides lactose depends on the availability of the oxygen.

Hetero lactic fermentation

The hetero fermentive bacteria lack some key enzymes that are necessary for the breakdown of fructose biphosphate. These enzymes are aldolase and triosephosphate isomerase. The initial breakdown of glucose therefore is entirely along the pentose phosphate pathway i.e., via glucose - 6 - phosphate —— 6 - phosphoclugonate —— ribulose - 5 - phosphate. Ribulose - 5 - phosphate is further converted into xylulose - 5 - phosphate by the enzyme epimerase. This pentose sugar is cleaved by a TPP dependent reaction catalysed by the enzyme phosophoketolase resulting in the formation of glyceroldehyde - 3 phosphate and acetyl phosphate.

Lactobacteria such as *Leuconostoc mesenteroides* ferment glucose to yield lactate, ethanol and carbondioxide. Obviously, they reduce the acetyl phosphate via acetyl - Co - A and acetyldehyde to ethanol. Many other heterofermentive bacteria convert the acetyphosphate either partially or completely to acetate, transfering the energy rich phosphate bond to ADP and thus producing ATP. The excess of hydrogen in this instance is transferred to glucose to give mannitol. The other compound glyceraldehyde - 3- phosphate is converted to lactate via pyruvate.

Fructose is also fermented by heterofermentive bacteria resulting in the formation of lactate, acetate, carbon dioxide and mannitol.

$$3 \text{ fructose} — \text{lactate} + \text{acetate} + CO_2 + 2 \text{ mannitol}$$

Another bacterium, *Lactobacillus plantarum* can ferment glucose homofermentatively, but breaks pentoses by the enzyme phosphoketalase to lactate and acetate. There are some bacteria such as *Lactobacillus casei,* which are homofermentive on glucose; become heterofermentive when the source substrate is ribose, and produce acetate and lactate.

Another important example of heterofermentive lactic acid bacteria is *Bifidobacterium bifidium,* which owes its name to the shape of its cells. This is predominantly present in the intestinal tract of breast fed babies. These bacteria depend heavily on sugars that contain N - acetylglucosamine, which is found only in human milk and not in any other milk. The species of *Bifidobacterium,* are strict anaerobes and require an atmosphere containing atleast 10% carbon dioxide. The metabolism of glucose by *Bifidobacterium* is as follows.

Glucose fermented via the phosphoketolase pathway. The microbe lacks aldolase and glucose - 6 - phosphagte dehydroganse, but has phosphoketolase. This enzyme can split fructose phosphate and xylulose - 6 - phosphate to acetylphosphate and erythrose - 4 - phosphate or glyceraldehyde - 3 - phosphate respectively.

$$2C_6H_{12}O_6 — 2CH_3\text{-}CHOH\text{-}COOH + 3CH_3COOH$$

Propionic acid fermentation

The bacteria that breakdown glucose to propionic acid are called *Propionibacteria.* Generally, these bacteria are present in the intestine of ruminants such as cattle and sheep where they play an important part in the formation of fatty acids particularly propionic and acitic acids. These bacteria are responsible for the conversion of lactate to propionate. *Propionibacteria* rarely occur in milk nor are they found in soil or water. They can be isolated by enrichment cultures under anaerobic conditions in a lactate yeast extract medium inoculated with Swiss cheese.

The following is a list of propionic acid fermenting bacteria.

Propionibacterium freudenreichii
P.acidi - propionici
P.acnes
Veillonella alcalescens (Micrococcus lactilyticus)
Clostridium propionoicum
Selenomonas
Micromonospora

Character of Propionibacterium

Cells are gram positive, non-motile rods and also non-spore forming. Propioni bacteria are strict anaerobes and are intolerant to atmospheric oxygen. Because of its intolerance, they are long regarded as obligatory fermentors. Some recent cytochromes and catalase. This suggests that under conditions of regulated aeration the cell growth perhaps will be better than under anaerobic conditions.

Hence propionibacteria are regarded as micro aerotolerant.

Under anaerobic conditions, propionibacteria ferment a variety of sugars such as glucose, sucrose, lactose and pentoses to porpionic acid. Even glycerol can be converted to propionic acid.

Wood and Werkman (1936), who have made a detailed study of glycerol fermentation by Propionibacterium acidi - propionici found out that the fermentation involves the net fixation of carbon dioxide. The fixed carbon dioxide was found in the excrete succinate. Carboxylation of pyruvate here is called, the Wood Werkman reaction.

$$3CH_3 - CHOH - COOH \longrightarrow 2CH_3 - CH_2 - COOH + CH_3 - COOH + CO_2 + H_2O$$

Reduction of lactate or pyruvate to propionate follows a pathway called methylmalonyl - Co - A pathway because of the characteristic intermediate compound methylmalonyl - Co - A. Pyruvate is carboxylated to oxalocetate by the enzyme methylmalonyl - Co - A. carboxytransferase. Oxalocetate is then reduced to succinate, this step is linked to an electron transport chain. The succinate gets converted to succinyl - Co - A and then gets transformed into methylmalonyl - Co - A by the enzyme metholmalonyl - Co - A mutase. This intermediate compound undergoes dcecarboxylation to yield propionyl - Co - A. Propionate is liberated from propional Co - A to succinate. Most of the propioni bacteria form propionate by the above mentioned pathway. The same pathway is used by Veillonella alcalescens and Selenomonas ruminatium also.

The methylmenloly - Co - A pathway is reversible and can be used to produce succinly - Co-.A.

Acryloly - Co - A pathway

Bacteria such as Clostridium propionicium, Bacteroides ruminicola and Megasphera elsedeni produce propionic acid by a simple route with derivatives such as acrylic acid, acryloyl - Co -A acting as the intermediate compound. This results in the formation of propionil - Co -A leads to the formation of propionate.

Formic acid fermentation

Some acid producing fermenting bacteria are collected together in a physiological grouping whose characteristic feature is the production of formic acid. In these bacteria, besides formic acid several other acids are also formed. Systematically, most of these bacteria are included in the family enterobacteriaceae. Formic acid producing interbacteria are gram negative, peritrichous, high motile and non spore forming rods. Physiologically they are facultatively aerobic and can obtain the energy either by aerobic, respiration or by fermentation.

Some of the common enterbacteriacease which produce formic acid are *Escherichia coli, Proteus, Enterobacter, Bifidiobacterium etc.*

Fermentation products

It has been pointed out early, that the fermic acid producing bacteria are also called mixed acid producing bacteria as they produce several acids in addition to formic acid. Many of the enterobacteriaceae bacillus species produces a number of organic products such as acetate, formate, succinate, lactate, ethanol, glycerol, acetoin etc. Sugars are largely metabolized via the fructose 1.6 diphosphate route and to a lesser extent by the pentose phosphate pathway.

Fermentation of glucose by Escherichia coli is characterized by the following reactions (Schlegel, 1986).

1. Conversion of pyruvate to acety Co - A and formate
2. Cleavage of formate to carbon dioxide and hydrogen
3. Reduction of acety Co - A to ethanol

The cleavage of pyruvate to acetyl Co - A and formate occurs only under anaerobic conditions and is catalysed by the enzyme pyruvate, formatelyase. This enzyme is very sensitive to oxygen. It is maintained in the reduced state by flavodoxin and requires for its activation, S - adenysyl - L - methionine.

Formate is split into carbon dioxide and hydrogen by most strains of C.coli as well as by other gas forming members of enterbacteriaceae. This reaction is catalyzed by the enzyme system formate: hydrogen lyase. The reaction is as follows :

$HCOOH \text{———} H_2 + CO_2$

The above reaction most probably involves the combined effects of a formate dehydrogenase and a hydrogenase

1. $HCOOH + X \text{——} CO_2 + XH_2$
2. $XH_2 \text{———} X + H_2$

Butyric acid - Butanol fermentation

Butyric acid, butanol, 2 - propanol and few other organic acids are produced as end products of carbohydrate fermentation by anaerobic bacteria belonging to the genus clostridium.

The genus clostridium belongs to the family Bacillaceae. The members are gram positive, motile (petritrichous), spore forming rods. Physiologically, clostridia are characterized by their intense fermentive metabolism. Species of clostridium, however, range from strict anaerobes (C.pasteurinum) to aerotolerants (C.acetobutylicum).

For their fermenting activity, clostridia use a variety of substrates such as polysaccharides, nuclei acids, proteins, purines, pyrimidnes etc.

Biochemistry of fermentation

Clostridal fermentation produces varying amounts of acids such as butryates, acetate, lactate etc. as well as alcohols (butanol, ethanol etc). In some instances, the end product of metabolism may even be acetone and gas (CO_2 or H_2.)

Clostridia metabolise glucose via the glycolytic pathway. Hydrogen removed from glyceraldehyde 3. Phosphate is added to organic acids synthesized from pyruvate and acetyl Co - A. Glucose fermentation by C. Butyrium and C. Acetobutylicum represents fermentation typical of the genus.

Butyrate is the product of condensation of two molecules of acetyl Co - A to acetoacetyl Co - A and subsequence reduction. Condensation of molecules is catalysed by the enzyme thiolase. Acetoacetyl Co - A is reduced by β-hydroxy butyryl - Co - A dehydrogenase and $NADH_2$ β-hydroxy butyryl Co - A. The enzyme corotonase removes water from this to give crotonyl CoA. This is then reduced by butyryl CoA dehydrogenase to butyryl CoA. The CoA can be transfered to acetate by CoA transferase to produce butyrate.

In a purely butyric acid fermentation, the hydrogen derived from the oxidation of pyruvate is liberated in gaseous form. When glucose is fermented to butyrate, CO_2 and H_2, a hydrogen balance is established and 3 mol ATP is regenerated per mol of glucose.

The end products of fermentation in Clostridium acetobutylicum are, butanol, butyrate, acetone and 2 - propanol. Initially, butyrate is excreted; with the increasing acidity, however acetoacelate decarboxylase enzyme becomes induced whose activity results in the formation of acetone and butanol. To be reduced to butano, butyrate must first get converted to butyryl CoA.

Fermentation of glutamic acid

Fermentation of glutamic acid attracted the attention of microbiologists, mainly because of the use of monosodium glutamate as a flavour increasing agent in food industry.

Japanese scientists devised a single stage fermentation to produce glutamic acid from strach (potato), using Micrococcus and Brevibacterium.

Presently, however, commercial production of glutamic acid is a two stage operation. In the first stage α- Ketoglutaric acid is produced by carbohydrate metabolism. A wide variety of bacteria produce α-Ketoglutaric acid (Pseudomonas fluroscens, Bacterium ketoglutarium etc). α Ketoglutaric acid is then converted to glutamic acid by the process of transamination.
E.coli and B. Ketoglutaricum produce glutamic acid with alanine or aspartic acid acting as amino group donor. The reactions take place as follows :
α Ketoglutaric acid + Aspartic acid —— glutamic acid + Oxalo acetic acid
Homoacetate fermentation

Bacterial activity that metabolizes substrate and ultimately proudces predominantly acetate is called homoacetate fermentation. Many of the species of Clostridia (C.formicoaceticum, C.thermoaceticum, C.acidicum etc), are able to transfer the hydrogen equivalents liberated in the initial oxidation of substrate only to carbon dioxide producing acetate.

$$8\,[H] + 2CO_2 \longrightarrow CH_3 - COOH + 2H_2O$$

C. Thermoaceticum and C.Formicoaceticum ferment glucose along the glycolytic pathway and produce almost 3 mol acetate per mol of glucose. Carbon dioxide released due to the decarboxylation of pyruvate gets refixed, and serves as H_2 acceptor. Pyruvate gives rise to acetate, CO_2, FdH_2 and ATP by the enzymes ferrodoxin oxidioreductase, phsophotransacetylase and acetate kinase.

MICROBIAL METABOLISM AEROBIC RESPIRATION

In the previous pages, anaerobic breakdown of carbohydrates has been studied ultimately leading to the formation of incompletely oxidised carbon compounds along with some amount of energy. In this chapter, we will study the aerobic breakdown of pyruvic acid leading to the complete oxidation of organic carbon and release of a large amount of energy. The essential feature of aerobic respiration is the participation of atmospheric oxygen.

Aerobic oxidation of pyruvic acid

This is the most important phase of aerobic respiration in which pyruvic acid gets converted into a two carbon compound called acetaldehyde, before entering the Krebs cycle for complete decarboxylation and oxidation. The reaction is mediated by a complex enzyme with atleast five essential cofactors (Gunsalnus 1954). The acetaldehyde combines with co enzyme A to form acetyl CoA which enters the Krebs cycle.

Formation of Acetyl Co A

As already pointed out, the enzyme which catalyzes the decarboxylation of pyruvic acids a complex one which includes 3 enzymes - *pyruvic acid decarboxylase, dihydroxylipoyl, transacetylase* and *dihydrolipoly dehydrogenase.* The enzyme complex also requires five cofactors – Thiamine pyrophosphate (TTP), Mg ions, NAD +, Coenzyme A (CoA) and lipoic acid.

The first step includes the formation of a complex between TTP and pyruvate, followed by the decarboxylation of the latter. In the second step, the acetaldehyde fragment reacts with the cofactor lipoic acid to form an *acetyl lipoic acid complex.* In this reaction, acetaldehyde is oxidized to the acid and lipoic acid is reduced. In the third step, acetyl group is removed from lipoic acid and added on to CoA, resulting in the formation of Acetyl CoA and reduced lipoic acid. In the final step, oxidized lipoic acid is regenerated by transforming the electrons from reduced lipoic acid to NAD +. This is necessary because, there should be a continuous supply of oxidized lipoic acid to accept pyruvate and produce acetyl CoA. The NAD which gets reduced to foom NADH + H^+ eventually enters into the electron transport system to produce 3 molecules of ATP. The following reactions summarize the formation of Acetyl CoA from pyruvate.

$$\text{(i) Pyruvate + TPP} \xrightarrow{Mg^{++}} \text{TPP Complex} + Co_2$$

$$\text{(ii) TPP Complex + Lipoic acid} \longrightarrow \text{Acetyl lipoic acid}$$

(iii) Acetyl lipoic acid + CoA ——— Acetyl lipoic acid (oxidized) complex; Complex + TPP, Acetyl lipoic acid (reduced form)

(iv) Lipoic acid + NAD ———— Lipoic acid + NADH + H^+
(reduced form) (oxidized form)

The net result may be given by the following equation.

Pyruvate + CoA + NAD Acetyl CoA + CO_2 + NADH + H^+

Finally, the two carbon compound Acetyl CoA formed in cytoplasm enters the mitochondria to react with the acids of the 'Kreb's cycle for the removal of remaining 2 carbon and to produce energy to be stored in the molecules of ATP.

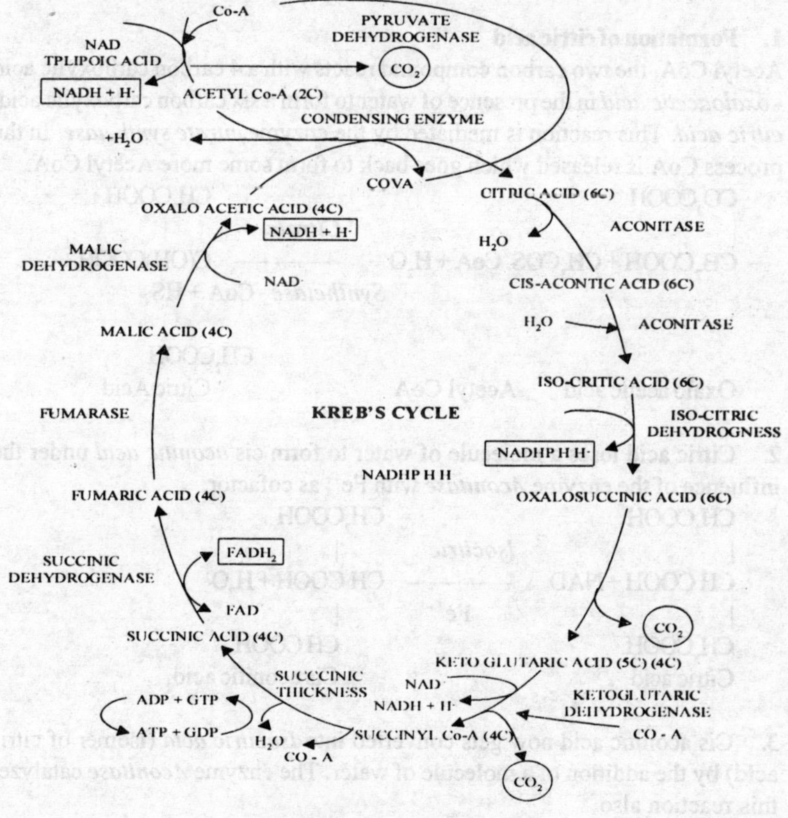

Fig. 6.4 Respiration
Reactions of Kreb's cycle

KREBS CYCLE

Acetyl CoA, the two carbon compound undergoes further decarboxylation when it enters a cycle of interconverting carboxylic acids present in mitochondria. Together with decarboxylation, there is oxidation also (dehydrogenation), resulting in the formation of molecules of ATP (to trap the released energy).

Krebs cycle (named after its discoverer), was actually worked out by Wood et al (1942), and Krebs (1943). The cycle has several names – Tricarboxylic XCycle (TCA), Citric acid Cycle, Mitochondrial respiration etc. Basically, the cycle consists of several 6 carbon, 5 carbon and 4 carbon acids. When these are getting interconverted, Acetyl CoA enters facilitating the interconversion, and in the process undergoes decarboxylation at two steps. The details of individual reactions are given below.

1. Formation of citric acid

Acetyl CoA, the two carbon compound reacts with a 4 carbon carboxylic acid - *oxaloacetic acid* in the presence of water to form a six carbon carboxylic acid, *citric acid.* This reaction is mediated by the enzyme *citrate synthetase.* In the process CoA is released which goes back to form some more Acetyl CoA.

CO_2COOH
|
$CH_2COOH + CH_3COS.CoA + H_2O$ —— *Citrate* *Synthetase* —— CH_2COOH | $C(OH)COOH +$ $CoA + HS$ | CH_2COOH

Oxalo acetic acid   Acetyl CoA   Citric Acid

2. Citric acid loses a molecule of water to form cis *aconitic acid* under the influence of the enzyme *Aconitase* with Fe^{++} as cofactor.

CH_2COOH
|
$CH\,COOH + NAD$ —— *Isocitric* Fe^{++} —— CH_2COOH | $CH\,COOH + H_2O$ | $CH\,COOH$
|
CH_3COOH

Citric acid   Cis aconitic acid

3. Cis aconitic acid now gets converted into *Isocitric acid* (isomer of citric acid) by the addition of a molecule of water. The enzyme *Aconitase* catalyzes this reaction also

CH_2COOH *Aconitase* CH_2COOH
| |

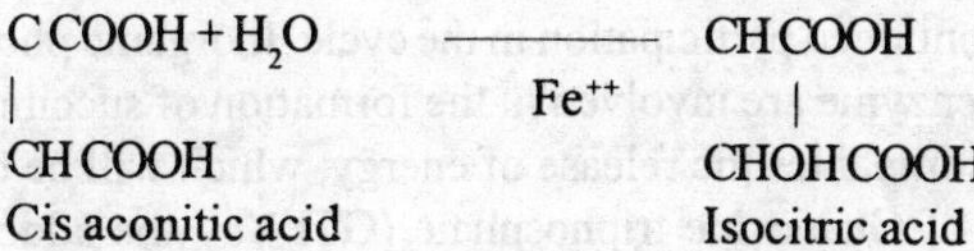

The addition of water molecule to cis aconitic does not producc Citric acid, but its isomer isocitric because H_2O combines in a different position in the molecule of a cis aconitic acid.

4. Isocitric acid undergoes oxidation (removal of hydrogen), under the influence of the enzyme Isocitric dehydrogenase to form *oxalosuccinic acid.* The released H atoms are accepted by NAD to become NADH + H +
The NADH + H + gets reoxidised after entering into the electron transport chain (see later).

CH2COOH		CH2COOH
|	*Isocitric*	|
CH COOH + NAD	————	CH COOH + NADH + H +
|	*dehydrogenase*	|
CHOH COOH		CO COOH
Isocitric acid		oxalosuccinic acid

5. Oxalo succinic undergoes decarboxylation to form a 5 carbon acid, α *ketoglutaric acid.* The reaction is catalysed by the enzyme *decarboxylase* with Mn^{++} as the cofactor.

CH_2COOH		
|	*Decarboxylase*	CH_2COOH
CH COOH	————	|
|	Mn^{++}	$CH_2CO\ COOH + CO_2$
CO COOH		

6. The next step is the oxidation of a α ketoglutaric acid which is achieved by the removal of H. Simultaneously, it also undergoes decarbosylation leading to the formation of Succinyl CoA. The enzymes involved are α *ketoglutaric dehydrogenase* and α *ketodecarboxylase,* together with coenzymes NAD and CoA. NAD gets reduced to $NADH + H^+$, while a molecule of CO_2 is liberated. Together with CoA, TPP, Mg^{++}, FAD and lipoic acid also participate in the reaction.

CH_2COOH	*Dehydrogenase*	CH_2COOH
| + HS.Coa + NAD	————	|
$CH_2\ CO\ COOH$	Mg^{++} TPP,	CH_2COS CoA
	Lipoic acid, FAD	Succinyl CoA
α Ketoglutaric acid		$NADH + H^+ + CO_2$

7. Succinly CoA is hydrolysed with a molecule of H_2O to *succinic acid,* regenerating CoA for continued participation in the cycle. Inorganic phosphate and a phosphorylating enzyme are involved in the formation of succinic acid. The release of CoA accompanies the release of energy, which will be used to synthesize a molecule of Guanosine triphosphate (GTP) by the addition of energy rich phosphate to Guanosine diphosphate (GDP). This is the only place where GTP instead of ATP is synthesized. This type of phosphorylation is called **substrate level phosphorylation.** The molecule of GTP subsequently reacts enzymatically with ADP to produce ATP and GDP.

(i) Succinyl Co.A + ip Sukccinyl-P + Co.A
(ii) Sukccinyl 1-P + GDP Succinic acid - GTP
(iii) GTP + ADP GDP + ATP

The enzyme succinate thiokinase catalyzes the reaction of conversion of succinyl CoA to succinic acid. From this reaction a molecule of ATP is synthesized.

CH_2COOH | + GDP + Pi CH_2CO - S.CoA —— *Succinate Dehydrogenase* ——→ CH_2COOH | CH_2COOH + GTP + HS.CoA

Succinyl CoA Succinic acid.

8. Succinic acid undergoes dehydrogenation in the presence of a flavoprotein enzyme – succinic acid dehydrogenase, to form fumaric acid. The pair of hydrogen atoms released are accepted by FAD (this is the only reaction in Krebs cycle where hydrogen is not accepted by NAD) and it gets reduced to $FADH_2$.s

CH_2COOH | + FAD CH_2COOH —— *Succinic acid Dehydrogenase* ——→ CH COOH | + $FADH_2$ CH COOH

Succinic acid Fumaric acid

9. One molecule of water is added to Fumaric acid enzymatically (Fumarase) to produce *malic acid.*

CHCOOH | + GDP + Pi CHCOOH —— *Fumarase* ——→ CHOH COOH | CH_2COOH

Fumaric acid Malic acid

10. In the last oxidation reaction of Krebs cycle malic acid undergoes dehydrogenation under the influence of *malate dehydrogenase* to prodeuce *oxalo acetic acid,* which combines with acetyl CoA to produce citric acid and continues the cycle. The hydrogen released from malic acid is accepted by NAD to produce NADH + H^+

$$\begin{array}{l} CHOH\,COOH \\ | \\ CH_2COOH \end{array} + GDP + Pi \xrightarrow[\textit{Dehydrogenase}]{\textit{Malate}} \begin{array}{l} CH\,COOH \\ | \\ CH_2COOH \end{array}$$

Malic acid → Oxaloacetic acid.

Summary of Krebs Cycle

In the Krebs cycle, there are four 6 carbon acids – *citric, cisaconitic, isocitric* and *oxalosuccinic* acids, one 5 carbon acid α *Ketoglutaric acid* and four 4 carbon acids *succinic, fumaric, malic* and *oxaloacetic* acids. When the 2 carbon acetyl CoA enters the cycle decarboxylation takes place at two stages – once to form the α Ketoglutaric acid and again to form succinic acid. With this, all the carbon atoms of the original glucose molecule have been removed.

The reactions of the Krebs cycle are for one pyruvate; since two molecules of pyruvate (3C) are formed at the end of glycolysis, the reactions of the Krebs cycle are to be doubled to account for complete glucose. The removal of CO_2 from glucose may be summarized as follows

Glycolysis

$$CH_6H_{12}O \longrightarrow 2CH_3CO\,COOH$$

Glucose → Pyruvic acid

In the above, there is no decarboxylation as all six carbon atoms are found in the organic molecule of pyruvate.

2CH3CO COOH (Pyruvic acid) → 2CH3CHO + 2CO2 (Acetaldehyde)

In the above, there is removal of two carbon atoms. Four more carbon atoms are left. They are removed as follows in the Krebs Cycle.

(i) 2 Molecules of Oxalosuccinic acid (6C) —— 2 molecules of α Ketoglutaric acid + $2CO_2$ (5C)

(ii) 2 Molecules of α Ketoglutaric acid (4C) —— 2 molecules of succinic acid (4C) + $2CO_2$

Thus in the aerobic respiration which involves Krebs Cycle, there is complete removal of all the organic carbon atoms to inorganic CO_2.

Another important aspect of Krebs Cycle is the dehydrogenation. The energy rich electrons released from the carboxylic acids at the following stages enter into the electron transport system (ETS) to produce molecules of ATP.

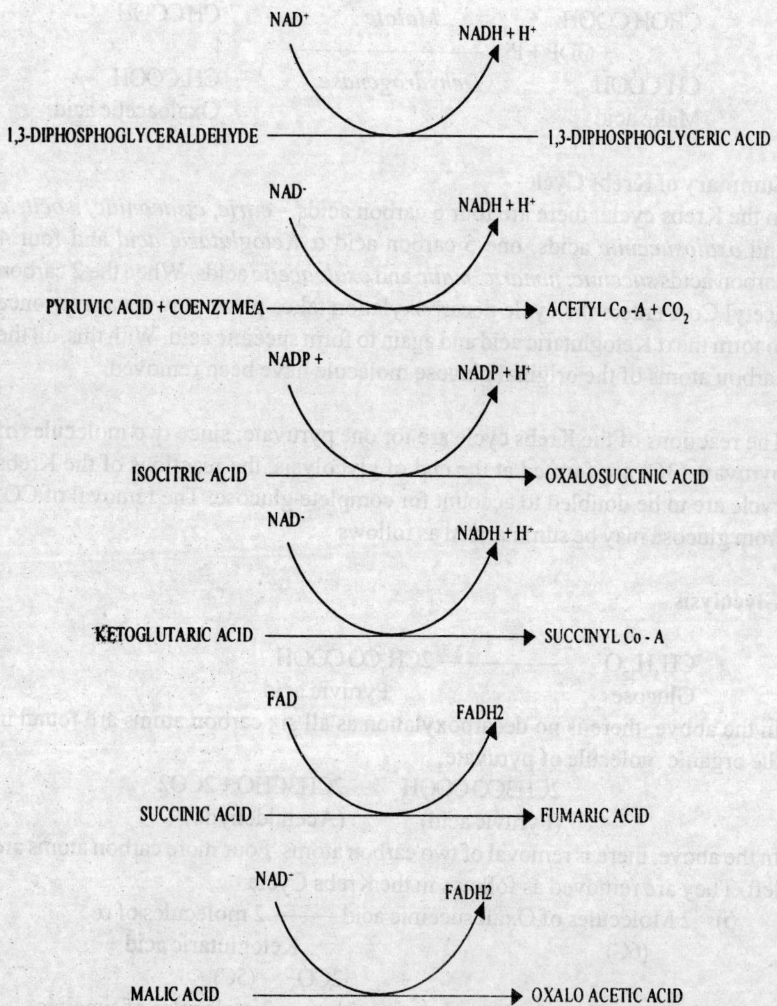

Fig. 6.5 Respiration

Dehydrogenation reactions in aerobic respirations (various stages)

Stages of Dehydrogenation in the Kreb's Cycle

(i) Conversion of Isocitric acid to oxalosuccinic acid

(ii) Conversion of α Ketoglutaric acid to succinyl CoA

(iii) Conversion of succinic acid to Fumaric acid

(iv) Conversion of malic acid to oxaloacetic acid

These are the stages at which the energy is trapped for the synthesis of ATP molecules. Besides the above, during the conversion of succinyl CoA to succinic acid there is substrate level phosphorylation (not through the ETS). All these are explained below in detail.

ELECTRON TRANSPORT SYSTEM (ETS)

Electron transport system or oxidative phosphorylation is the final stage in aerobic respiration. Oxidative phosphorylation takes place in the F particle or oxysomes of the cristae in mitochondria.

In various steps in the respiratory process, the substrates are dehydrogenated (oxidised), and a pair of hydrogen atoms are removed. These H atoms are accepted by a series of acceptors which in turn get reduced and oxidised alternately. As hydrogen is shuttled between acceptors it ultimately reaches the last acceptor and then combines with 1/2 molecule of O_2 (O) to form water. The series of acceptors constitute the ETS.

The bonded hydrogen cannot combine with atmospheric oxygen to form water. In ETS hydrogen is released from its bonded state. The H atoms get disassociated into two protons and two electrons ($2H^+ + 2e^-$). The two protons released in the medium as H^+ ions are available for combination with oxygen to form water. At each dehydrogenation, two atoms of hydrogen are released with the result at every step, after disassociation four protons ($4H^+$), and four electrons ($4e^-$) are released. While the H_+ ions are used for the formation of water, electrons, more specificaly the energy in them, is utilized for the formation of molecules of ATP.

Components of ETS

The paris of hydrogen ($2H^+ + 2e^-$) removed from the substrate at various steps are eventually transported to the oxygen through an assembly of enzymes which constitute the respiratory chain. The enzymes of the respiratory chain mediate in the transport of the hydrogen through their prosthetic groups or coenzymes. The prosthetic group of an enzyme in the respiratory chain undergoes reduction when it accepts hydrogen or electrons from the preceeding enzyme, and passes it on to the next enzyme in the chain getting oxidized (coming back to normal, in the process.

It is believed that most of the enzymes which are involved in dehydrogenating the substrate are loosely bound to the mitochondrial inner membrane, while the others (enzymes) constitute the integral part of the lipoprotein matrix of the inner membrane.

The ETS has the following enzyme complex and two mobile carriers.

Complex I

This consists of a flavoprotein with attached $NADH_2$ dehydrogenase enzyme having FMN (Flavin mononucleotide) as the prosthetic group. 2–4 atoms of nonheme iron with iron-sulphur (Fe-S centre, is found in the enzyme.

Complex II

This also consists of a flavoprotein called *succinic dehydrogenase* with FAD (Flavin adenise (dinucleotide) as its prosthetic group. In addition, there are two Fe-S centres an i-e- S protein. Between Complex I and II, is present a mobile carrier ubiquinone (CoQ). CoQ is reduced by $NADH_2$ as well as succinic hydrogenase.

Complex III

This consists of Cyhtochrome *B* and C_1 with a non heme iron protein with only one Fe-S centre. However, the presence of Cytochrome C_1, is doubtful in plant ETS.

Complex IV

This is also called cytochrome oxidase or terminal oxidase. The complex consists of cytochromes *a* and a_3, having one or two atoms of firmly attached copper. Another mobile carrier is found between Complex III and IV which has a prosthetic group made up of iron prophyrin IX. This is identical with the heme of hemoglobin.

Complex V

This is the enzyme system called ATP ase and is mainly concerned with the synthesis of ATP from ADP and inorganic phosphate.

The plant electron transport system consists of these complexes and their location and sequence may be decided by the evidence provided by redox potential and respiratory inhibitors.

Evidence from Redox potential

In the biological system, certain compounds more readily get oxidized (give out electrons), while others get reduced (accept electrons). This phenomenon of release or acceptance of electrons is called **Redox potential** (E'ϕ). It is usually defined as the ratio between the reduced and oxidized forms of a compound. The standard electrode potential (Eo) may be defined as the potential at 50% oxidation for an oxidation reduction system at pH7. The standard electrode potential measurement (measured in milivolts) of the various components of the ETS gives the following sequence (in the order of decreasing potential), with exception of CoQ. The sequence is NAD, FMN, FAD, *Cytb, Cytc, Cyte, Cyta,* and *Cyta3.* The standard redox potential of NADH2 is 0.32 volts while that of water is + 0.82 volts. It is for this reason that electrons from hydrogen to oxygen will yield energy.

Evidence from respiratory inhibitors

The correct sequence of the components of ETS can also be ascertained by the use of specific inhibitors, which block the function of one or the other components. For instance, when dilute cyanides are used which block the functioning of *Cyta* and *a*3, all other components get reduced i.e., function normally indicating *Cyta* and *a*3 are at the end. If, on the other hand antimycin A is used, it inhibits the functioning of NAD, all other components remain in the oxidized state (will not get reduced), thus indicating that NAD must be the first electron acceptor.

Stages of dehydrogenation

Total oxidation of the substrate material (glucose) into CO_2 and water involves several dehydrogenations at various steps (in glycolysis and Krebs cycle). The released hydrogen is picked up by the NAD or FAD and passed on to the next electron acceptor.

1. 2H is removed in glycolysis when a molecule of glyceraldehyde 3. Phosphate is converted into 1.3 Diphosphoglyceric acid. NAD picks up the hydrogen and gets reduced to $NADH_2$.
2. During the decarboxylation and dehydrogenation of pyruvic acid into acetyl CoA. 2H is removed. Hydrogen is accepted by NAD to become NADH2.
3. 2H are removed from isocitric acid to form oxalosuccinic acid. NAD accepts hydrogen and is reduced to $NADH_2$.
4. During the oxidation of α Ketoglutaric acid to succinic and 2H are removed. This is again accepted by NAD to become $NADH_2$.
5. During the formation of fumaric acid from succinic acid, dehydrogenation occurs which reduces FAD to $FADH_2$ (Note : This is the only instance where FAD, and not NAD is the first electron acceptor from the substrate).
6. During the dehydrogenation of malic acid to oxaloacetic acid, 2H atoms are removed which are picked up by NAD to get reduced to $NADH_2$.

Thus in the above 6 steps (one in glycolysis, one in pyruvate and four in Krebs cycle), a total of 12 hydrogen atoms are removed. Since this number is for triose only, it has to be doubled for a hexose and it gives a number of 24 hydrogen atoms. These hydrogen atoms travel down their electrical gradient along the various electron acceptors, which alternately get reduced and oxidized as the electron in shuttled through. At specific stages, the energy in the electron in made use of for the synthesis of ATP (oxidative phosphorylation).

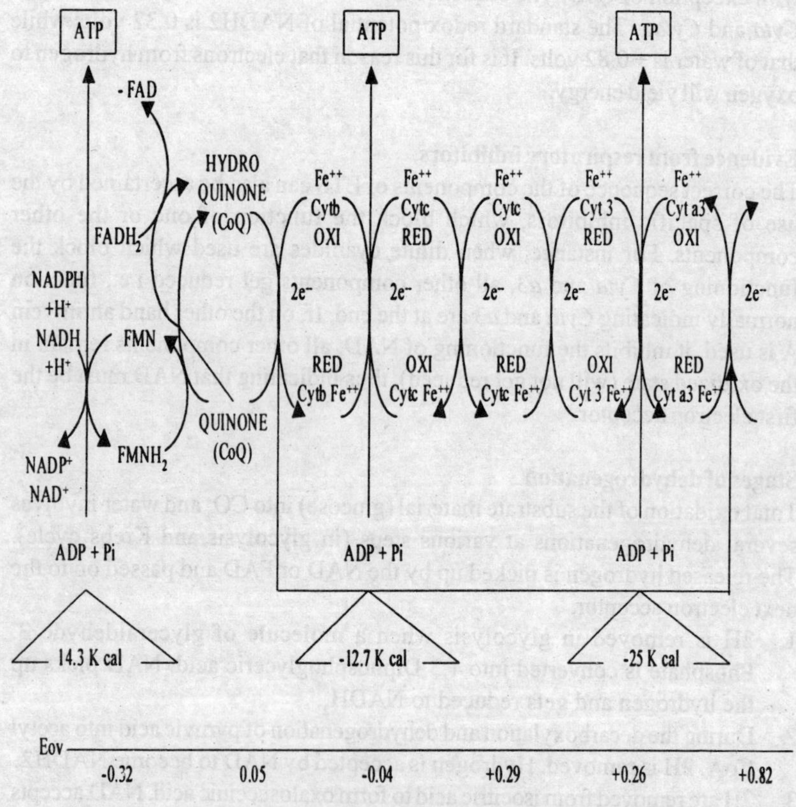

Fig. 6.6 Respiration

Schematic representation of electron transport system in oxidative phosphorylation

The Scheme of Electron Movement

The ETS begins with the acceptance of 2 atoms of hydrogen from the substrate by NAD which gets reduced to $NADH_2$. $NADH_2$ is oxidized when the 2H are transfered to the prosthetic group os lipoproteins of the FMN which in turn gets reduced to $FMNH_2$.

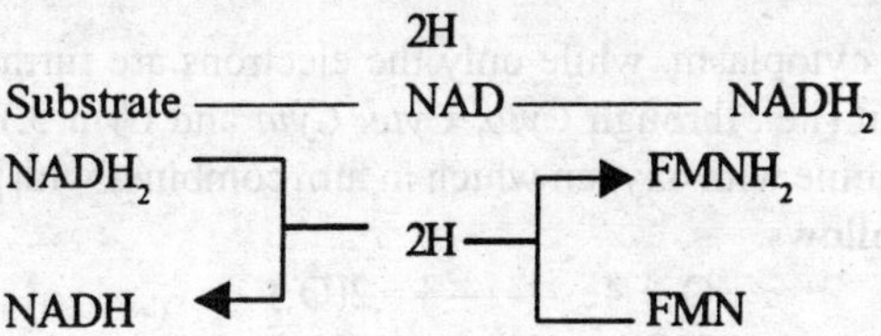

The 2H atoms are now transferred from $FMNH_2$ to the mobile carrier CoQ (Coenzyme Q or uniquinone). CoQ is chemically related to vitamin K and E and similar to plastoquinone of chloroplasts. This CoQ links the flavoproteins of the FAD to Cytochrome.

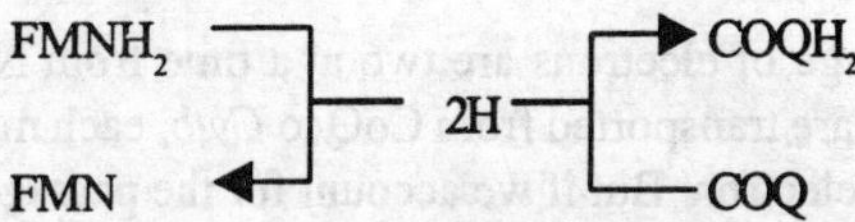

For hydrogen removed in all instances, NAD is the initial acceptor except in the case of succinic acid (Krebs Cycle). Here, the hydrogen is picked up by FAD (Flavin adenine dinucleotide), which is reduced as follows.

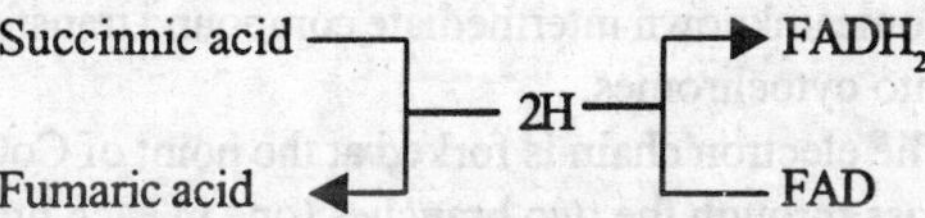

The reduced FAD ($FADH_2$), then passes on its hydrogen to CoQ or *Cytb*. The actual point of hydrogen entry (either to CoQ or *Cytb*) is not yet difinitely established.

Some of the current researches have indicated that at two places (pyruvate and α Ketoglutrate), hydrogen released from the substrate does not go direct to NAD but through lipoate and FAD and then to NAD. This is, however, subject to further proof.

According to Kligenberg and Kroger (1967), CoQ not only picks up hydrogen from succinate *via* FAD but also from α glycerophosphate and Acetyl CoA. The FAD of acetyl CoA, an additional flavoprotein ETF (electron transporting flavoprotein) is present.

The reduced CoQ is oxidized by the released of 2 hydrogen atoms. Soon after release, hydrogen undergoes disassociation into two, protons ($2\ H^+$) are released into cytoplasm, while only the electrons are further transported through the chain (i.e., through *Cytb, Cyte, Cyta* and *Cyta* 3.) The two electrons finally combine with oxygen which in turn combines with protons (H+) to form water as follows.

$$O_2 + 4e^- \longrightarrow 2(O_2)$$
$$2(O_2) + 4H^+ \longrightarrow 2H_2O$$

(In the above equation, $4e^-$ and $4H^+$ are shown instead of $2H^+$ and $2e^-$. We have to remember that the reaction sequences of glycolysis (from 3 PGA onwards), and Krebs Cycle are for only half of glucose (one triose), hence they have to be doubled while accounting for a complete glucose molecule).

The passage of electrons are two at a time from NAD upto CoQ. But when electrons are transported from CoQ to *Cytb,* each molecule of *Cytb* can accept only one electron. But if we account for the passage of only one electron at a time, it will not be sufficient to reduce molecular oxygen. How the two electron chain is changed to a single electron chain (at the same time delivering two electrons towards the end), is as yet completely not known. The following are some of the views.

(a) Some unknown intermediate compound transfers the electrons one by one into cytochromes.
(b) The electron chain is forked at the point of CoQ so that two electrons can pass through the two branches (one in each branch). Each branch has the same enzyme system as the other. Finally, the two electrons released from the branches combine with molecular oxygen.

(c) The electron chain is formed at the point of CoQ so that the two electrons can pass through the two branches (one in each branch). Each branch has the same enzyme system as the other. Finally, the two electrons released from the branches combine with molecular oxygen.

The third view seems to be possible as it satisfactorily accounts for the smooth flow of both the electrons in separate but parallel channels.

The electrons give out by CoQ are first accepted by Cytochrome *b* which gets reduced. *Cytb* is reoxidised by transferring its electron to *Cyt c.* From these, the electron passes to *Cyt c, Cyt a,* and finally to *Cyt a*3. *Cyt a*3 gets oxidized by transferring electron to the atmospheric oxygen. How is it released? Each enzyme consists of a heme prosthetic group called *protoheme.* This iron alternately gets reduced and oxidized by accepting and donating an electron at a time as follows.

Cyt b CFe^{2+} Cyt*c* (Fe^{2+})

e

Cyt b (Fe^{3+}) $Cytc_1$ (Fe^{3+})

Some physiologists argue that *Cyt* cl is not a component of plant ETS. Hence, thcy believe that *Cyt h* transfers the electrons to *Cytc*.

ATP synthesis (oxidative phosphorylation)

All biological reactions require energy and this depends on the formation and utilization of, energy rich' phosphate bonds. A high energy bond does not actually mean that the bond is strong, on the contrary the bond is labile and when it is broken in a reaction, a large amount of energy is released. The energy requiring reactions always obtain energy by using certain compounds which readily release energy. These are called energy currencies. Similarly, during the storing of energy that is released in a reaction, these intermediate molecules are synthesized. Several molecules which act as energy intermediaries are known in biological systems. Of these, the most important are ATP (Adenosine Triphosphate), and ADP (Adenosine Diphosphate). Both ATP and ADP are derived from AMP (Adenosine Monophosphate). AMP consists of a purine (Adenine), a pentose sug (ribose), and phosphoric acid. ADP and ATP are similar to AMP except in the number of phosphate groups. While ADP has two phosphate groups, ATP has three. The second and third phosphate bonds are energy rich in ATP. On hydrolysis of the phosphate bond, energy equivalent to 7600 cal per mol is released.

To sum up, whereeever energy is released, ATP molecules are synthesized and whenever energy is required ATP molecules are utilized. This may be illustrated by the following equations.

Synthesis

$$ADP + Pi \xrightarrow{\text{energy}} ATP$$

(inorganic phosphate)

Utilization

$$ATP + X \longrightarrow X\text{-}(P) + ADP$$

(X = Substrate that requires energy)

ATP molecules are synthesized in living systems by two methods.

(i) Photophosphorylation and

(ii) Oxidative phosphorylation.

In photophosphorylation (seen in green plants), radiant energy is used to add Pi to ADP to produce ATP. In oxidative phosphorylation, ATP is synthesized using the energy of oxidation.

Oxidative phosphorylation is the most significant functions of ETS. The number of ATPs produced may be explained on the basis of P/O index (phosphorylation/ Oxygen). P/O index is the ratio of ATP produced for each atom of oxygen used.

Observations of Lehninger point out that the P/O ratio in the case of NAD linked ETS is three *i.e.,* at three sites (between NAD and molecular oxygen), question as to why there are only three energy producing sites, the answer may be provided as follows.

When the electron passes through the chain from one acceptor to the other, at every step some energy is released. The energy so released wil be equivalent to the difference in energy level between the electron donor and acceptor. At certain steps, the amount of energy released is sufficient to bind Pi to ADP, and hence a molecule of ATP is synthesized. As per some calculations, it has been noted that 0.28 volts of energy is sufficient to synthesize a molecule of ATP from ADP and Pi, and in the ETS, whereever the energy difference between the donor and acceptor is 0.28 volts or more, a molecule of ATP is synthesized. As has already been pointed out, in substrate systems linked to NAD there are three sites, where a molecule of ATP is synthesized.

These are

(i) NADH	FAD
(ii) *cyt b*	*Cyt c* or *c,* and
(iii) *Cyt a*	$Cyta_3$

In the oxidation of succinnic acid which is linked to FAD directly, obviously there are only two sites for the synthesis of ATP, hence only two molecules of ATP are synthesized.

The scheme of ETS and the sites of ATP molecule synthesis proposed by Bidwell (1979), are slightly different, though he also agrees that in NAD linked substrate systems there are three sites of ATP synthesis and in FAD linked systems there are only two. But their places the CoQ (Ubiqinone) between *Cyt b, Cyt c* and not earlier to the cytochrome systems (between FAD and Cytochrome b). This poses a problem for the movement of electrons as ubiquione requires 2 H^+ atoms for reduction and not just electrons. But before the 2H reaches the Cytochrome b, it would have undergone disassociation with two protons ($2H^+$) getting released into cytoplasm. Hence, Bidwelo (1979), believes that once again (after ionization before they reach Cyt b) $2H^+$ and $2e^-$ unite to form H, and this reduces the ubiquinone. From ubiquinone when 2 H has to reach *Cyt c*, there is ionization, $2H^+$ is released into water while $2e^-$ continues in the ETS. Hence in this chain, there is ionization of hydrogen twice and not once as believed by other physiologists.

Mechanism of ATP formation

Same as in Phosophosphorylation.

Balance sheet of energy in Respiration

(Account of ATP molecules synthesized). The following is a total of ATP molecules synthesized during complete breakdown of glucose to CO_2 and H_2O

Glycolysis

	Reaction Sequence	ATP Used	ATP Synthesized
1.	Glucose to Glucose 6 Phosphate	1	-
2.	Fructose 6 Phosphate to Fructose - 1,6 - Diphosphat	1	-
3.	1.3 Diphosphoglyceric acid to 3 Phosphogyceric acid (1 x 2)	-	2
4.	Phosphoenol pyruvic acid to pyruvic acid (1 x 2)	-	2
5.	Dehydrogenation of Glyceraldehyde - 3 Phosphate to 1.3 Diphosphoglyceric acid (through ETS with participation of NAD) 3 x 2	-	6
	Total	**2**	**10**

Net gain of ATP during Glycolysis = 8 molecules.

(Note : Reactions 1 and 2 are for entire glucose molecule, while, 3,4 and 5 are only for a triose, hence they are doubled)

Pyruvic acid Oxidation

Reaction Sequence	ATP Used	ATP Synthesized
Conversion of pyruvic acid to Acetyl CoA; Dehydrogenation through ETS with initial acceptor being NAD (3 x 2)		6

Krebs Cycle

(There is no ATP loss in Krebs Cycle)

	ATP Synthesized
1. Isocitric acid to oxalosuccinic acid dehydrogenation through ETS with NAD participation (3 x 2)	6
2. Conversion of Ketoglutaric acid to succinyl CoA, ETS with NAD as the initial acceptor (3 x 2)	6
3. Conversion of succinyl CoA to succinic acid; substrate level phosphorylation	2
No ETS (1 x 2)	
4. Conversion of succinic acid to Fumaric acid ETS beginning with FAD and not Nad. Hence only two sites of ATP synthesis (2 x 2)	4
5. Conversion of malic acid to Fumaric acid. ETSbeginning with NAD (3 x 2)	6
Total	**24**

Total ATP used during the entire aerobic respiration = 2

Total ATP produced during glycolysis, pyruvic acid oxidation and Krebs Cycle. 10 + 6 + 24 = 40

Net gain of ATP molecule for aerobic oxidation of one molecule of glucose 40-2=38

Difference between photophosphorylation and oxidative phosphorylation

No.	Character	Photo-Phosphorylation	Oxidative Phosphorylation
1.	Occurrence	Photosynthesis	Respiration
2.	Cell organelle involved	Chloroplast	Mitochondrion
3.	Occurrence within the organelle	Thylakoid	Inner membrane Memgranes of cristae
4.	Need for molecular oxygen	Not needed	needed
5.	Source of energy	Light	Reduction-oxidation reactions
6.	Output of ATP molecules	Less	More
7.	Fate of ATP molecules	Used up for the Reduction of CO_2	Released into cytoplasm and available for all metabolic reactions.

Efficiency of Respiration

A molecule of glucose consists of a total amount of 6,86,000 (▲ F) calories of energy in the form of chemical bond. During the complete oxidation of a glucose molecule 6,74,000 (673.6 Kcal) calories of energy is released. Out of this, only a part is trapped in the energy rich phosphate bonds of ATP molecules, while the rest is released as heat. Each ATP molecule has about 7,600 calories of energy. The total amount of energy trapped during the breakdown of glucose depends on the number of ATP molecules formed.

In case of oxidative phosphorylation, the total number of ATP molecules produced will be 38. This means a total of 38 x 7600 x 100/6,86,000, or about 40% of the total energy will be negotiable, while the rest (60%) is wasted. In the alternate pathways of glucose breakdown (pentose phosphate shunt), a similar amount of energy will be made available in the form of molecules of ATP. This gives an efficiency of 40% for aerobic breakdown of glucose.

In anaerobic respiration, only two molecules of ATP are formed per glucose molecule. The efficiency will be

$$C_6H_{12}O_6 \longrightarrow 2C_3CH_4O_3 + 4H$$

(F = 52 Kcal)

= (2 x 7.6/52) x 100 = **29.2%**

The energy efficiency for alcoholic fermentation would be (here also only two ATP molecules are formed)

$$C_6H_{12}O_6 \longrightarrow 2C_2C_5OH + 2CO_2$$

(F = 52 Kcal)

= (2 x 7.6/52) x 100 = **29.2%**

OXIDATION OF INORGANIC COMPOUNDS

In chemolithotrophs, energy is obtained by oxidation of inorganic compounds. Chemolithotrophs are generally identified by their ability to grow in mineral media and to obtain their cellular carbon from carbon dioxide.

Chemolithotrophs may be classified into obligate chemolithotrophs and facultative chemolithotrophs. Obligate chemolithotrophs generally cannot utilise organic compounds as their carbon source. It was earlier believed that for these organisms organic compounds are toxic, but some of the recent studies have shown that organic compounds are not completely toxic eventhough they partially inhibit the growth of microorganisms. In facultative chemolithotrophs, the main source of cellular carbon is CO_2; they can use even organic compounds as the source of their CO_2 requirement.

Oxidation of inorganic compounds is the chief source of energy both for obligate and facultative chemolithotrophs. Inorganic substances like ammonia, nitrate, hydrogen, sulphur compounds, iron compounds etc. get oxidised and release to glucose. All chemolithotrophs can also be called autotrophs. In the biosphere, they are the chief non-photosynthetic primary producers. The following are some of the examples of oxidation of inorganic compounds.

1. Oxidation of nigtrogenous compounds

A. Ammonia Oxidation

This is seen in organisms such as *Nitrosomonas* and *Nitrosocystis.* Here, the electron donor is ammonium ion and the electron acceptor is oxygen. The oxidation of ammonium ion to nitrate provides the necessary energy for CO_2 fixation. During the process, water is a byproduct. The reaction takes place as follows :

$$NH_4 + 3/2O_2 \text{ —— } No_2 - H_2O + 2H^-$$

Two steps are involved in the oxidationo of ammonia to nitrate. In the first step, ammonia is converted into hydroxylamine involving the expenditure of energy. Between hydroxylamine and nitrate, there is atleast one intermediate compound. The electrons liberated by hydroxylamine pass through a series of electron acceptors such as falvin, cytochromes (b,ca) and produce ATP. An ATP dependent reverse electron flow helps in the reduction of NAD to NADH2. The reduction of NADH is apparently at the cost of ATP. The nature of the intermediate compound between hydroxylamine and nitrite is still uncertain.

The conversion of ammonia to nitrite is called nitrification. This process also involves two steps. In the first step, ammonia is converted into nitrate by bacteria such as *Nitrosomonas* and *Nitrosocystis.* In the second step, the nitrite is oxidised to nitrate by nitrite oxidising bacteria like *Nitrobacter.* Nitrification is of great importance in the maintenance of nitrogen balance in nature.

The conversion of ammonia to nitrate and the release of ATP takes place as shown in the following equation:

$$NH_4 + \text{ —— } NH_2OH \text{ —— Intermediate } NO_2$$

$$\downarrow e$$

$$\text{Flavin —— Cyt b —— Cyt c —— Cyt a —— } O_2$$

$$\downarrow \text{ATP (from Cyt c)}$$

B. Nitrate Oxidation

As has been mentioned above the bacterium involved in nitrite oxidation is *Nitrobacter*. This bacterium utilises nitrite as the source of electrons and oxygen as the acceptor of electron. Oxidation of nitrite to nitrate is achieved at one step as follows :

$NO_2 + 1/2\, O_2 \quad NO_3$

It has been found that Nitrobacter can obtain energy for its growth from the oxidation of formate. Enzyme such as nitrite oxidase and formate dehydrogenase are found in the same respiratory particle.

Oxidation of sulfur compounds

Many of the chemolithotrophs oxidise/reduce sulphur compounds to obtain their energy. These chemolithotrophs include sulphur bacteria both green and purple. The inorganic compounds that get oxidised are sulphide, thiosulphate, thiocyanate and elemental sulphur. *Thiobacillus* can oxidase sulphur and sulphur compounds such as sulphides and sulphites.

Oxidation of sulphur (S^0)

Many of the species of *Thiobacillus,* such as T.*Thiooxidans,* can convert both sulphur and sulphite to sulphite to sulphate. *T.Thiooxidans. T.Concretivorous* and *Thioparus* can aerobically oxidise elemental sulphur to yield sulphate. In the oxidation of sulphur to sulphate, sulphate is the intermediate compound. Experiments conducted with cell-free extracts of *T.Thiooxidans* shows that the elemental sulphur first reacts with oxygen to produce sulphite. This reaction is cartalysed by the enzyme oxygenase and is dependent on glutathione. The reaction is as follows:

$$SO_8 + 12O_2 + 8H_2O \longrightarrow 8H_2SO_3$$

The sulphite that is produced as a result of the above reaction noenzymatically reacts with elemental sulphur to yield thiosulphate as per the following reaction:

$$H_2SO_3 + SO \longrightarrow H_2S_2O_3$$

Oxidation of Sulphide

Some Thiobacilli are known to use sulphide. During the oxidation of sulphide to sulphate 8 electrons are released. Green and purple sulphur bacteria can anaerobically oxidise reduced sulphur compounds like H_2S. Both sulphide and elemental sulphur are metabolised due to the interaction of the enzyme bound thiol or dithiol. Sulphite and elemental sulphur react with a sulphydryl group to yield an intermediate RSSH.

This intermediate is oxidised to sulphite by sulphide oxidise. There is no generation of ATP here as the reaction is not linked to cytochrome electron transport. At the next step sulphite is oxidised to sulphate by the enzyme sulphite oxidase. As this reaction is cytochrome linked it can generate ATP through ADP.

Oxidation of sulphite to sulphate

Many of the thiobacilli are known to bring about oxidation of sulphite to sulphate. This reaction takes place through an AMP-dependent APS - (adenosine phosphorulphonate) reductase pathway.

APS is a high energy compound and it is formed due to the interaction of sulphite and AMP (Adenosine Monophosphate). This reaction is catalysed by the enzyme APS reductase.

$$2SO_3 \text{——} + 2\,AMP \xrightarrow{\text{APS reductase}} 2\,APS + 4e^-$$

At the same time inorganic phosphate is esterified to ADP, and the reaction is catalysed by the enzyme ADP sulphurlyase.

$$2APS + 2Pi \xrightarrow{\text{ADP sulphurlyase}} 2ADP + SO_4$$

ATP is now generated from ADP by the action of the enzyme adenylate kinase.

$$2\,ADP \longrightarrow AMP + ATP$$

The sum of the above reactions can be studied as follows :

$$2SO_3 \text{——} + AMP + 2Pi \longrightarrow 2SO_4 + \text{——} ATP + 4e^-$$

Iron oxidation

Some chemolithotrophs such as *Thiobacillus, Ferroxidans* can oxidase ferrous to ferric iron as follows :

$$4\,Fe_2^+ + 4H^+ + O_2 \longrightarrow 4\,Fe_3^+ + 2H_2O$$

As a result of the oxidation, electrons that are released will flow from ferrous compound to cyt c and cyt a, and ultimately react with moleculare oxygen. A molecule of ATP is formed between cyt a and O_2.

Oxidation of hydrogen

Hydrogen is produced during the anaerobic degradation of organic substances in soil. Many chemolithotrophic bacteria have the ability to utilise this hydrogen. In order to oxidis hydrogen, bacteria should possess enzymes such as hydrogenase. It has been seen that hydrogen oxidising bacteria have evolved independently and the enzyme hydrogenase occurs in at least three different forms –

1. Cannot reduce NAD and is inhibited by CO. This is the commenest one.
2. It is called the nitrogenase type and it reacts with ATP and Fd resulting in

the release of molecular hydrogen.

3. It is called the *Hydrogenomonas type,* it reduces NAD and is inhibited by CO.

Oxidation of hydrogen is of the following type:

1. Oxidation reduction
2. Nitrate reduction
3. Sulphate reduction
4. Methanogenesis

1. Oxidation reduction

This is seen in the bacterium *Hydrogenomonas.* Oxidative phosphorylation linked to the reduction of oxygen by molecular hydrogen provides energy necessary for growth.

$$H_2 + 1/2\ O_2 \longrightarrow H_2O$$

The energy released is used for the reduction of carbon dioxide to cellular carbon. Oxidation of hydrogen provides energy through which cellular synthetic activities take place. Hydrogenase enzyme can split hydrogen as follows :

$$H_2 \longrightarrow 2H^+ + 2e^-$$

The electrons released have a low electrifcal potential, hence the carriers also should have a low redox potential.

In *Hydrogenomonas,* the electron transfer chain takes place as follows :

H_2 —— Intermediate carrier —— Flavoprotein —— Cytochromes —— O_2

|

NAD

2. Nitrate reduction

This has been seen in bacteria commonly called denitrifying bacteria (Micrococcus denitrificans). These bacteria use nitrate as the terminal acceptor of electrons instead of oxygen. Nitrate is reduced with molecular hydrogen. The bacteria possess type 1 hydrogenase enzyme. As NAD cannot be reduced directly, an ATP dependent reduction system is used. The reduction of nitrate is believed to be a process occurring in several steps as follows:

$$NO_2 \longrightarrow NO_2 \longrightarrow NO \longrightarrow NO_2 \longrightarrow N$$

3. Sulphate reduction

Two types of sulphate reduction have been noticed in bacteria. These are *Assimilatory sulphate reduction,* and *Dissimilatory sulphate reduction.*

Assimilatory sulphate reduction

In this process, sulphate is reduced to sulphide which is used for the synthesis of sulphur containing amino acids like cysteine and methionine. This is called

Assimilatory sulphate reduction because the sulphate is assimilated into the amino acids. The following steps are involved in Assimilatory sulphate reduction.

1. Activation of sulphate by ATP. In this process, inorganic sulphate reacts with ATP and is catalysed by the enzyme ATP sulphurylase resulting in the formation of APS.
2. Phosphorelation of APS. APS is phosphorelated with the help of ATP to produce a compound called PAPS (3 phospho adenosine, 5 phospho sulphonate).
3. Reduction of sulphate. The sulphate in PAPS is reduced to sulphite by the action of the enzyme PAPS reductase. This reduction involves the transfer of electrons from NADH to PAPA.
4. Reduction of sulphite. NADPH reduces sulphite to sulphide under the influence of the enzyme sulphite reductase.

$$SO3 - 3NADPH + 3H^{+} \longrightarrow H_2S + N3NAD + + 3H_2O$$

The hydrogen sulphide released in the above reaction is immediately utilised for the formation of ammino acid cysteine. This reaction is catalysed by the enzyme sulphydrase.

Dissimilatory sulphate reduction

A few species of gram negative (*Desulfovibrio*), and gram positive (*Desulfotomaculum*) bacteria are capable of dissimilatory sulphate reduction. These bacteria are strictly anaerobic.

Dissimilatory sulphate reduction produces hydrogen sulphide which is found in the cytoplasm of bacteria such as *Desulphovibrio.* These bacteria can also use pyruvate, lactate etc. as substrates. *D.Desulfuricans* can also utilise molecular hydrogen for the reduction of sulphate as follows :

$$SO_4 -- + 4H_2 — S -- + 4H_2O$$

The first step in the reduction of the sulphate is the formation of APS. The reaction is similar to what is seen in the Assimilatory sulphate reduction. In the next step, APS is reduced to sulphite and AMP under the influence of the enzyme APS reductase. Further reduction of sulphide is brought about by the enzyme sulphite reductase. In the above process, Cyt c3 is an electron carrier.

4. Methanogenesis

In certain bacteria, oxidation of hydrogen takes place and carbon dioxide acts as a terminal electron acceptor. Reduction of carbon dioxide here results in the production of methane. Bacteria which reduce carbon dioxide to methane are called methanogenic bacteria or methanogenic bacteria or methanogens. Mathanogenic bacteria are found in the intestine of many animals and also in soil and in sewage tanks. Physiologically, they are strict anaerobes but Autotrophic, as they use only carbon dioxide as the source. A few examples of methanogenic bacteria are *Methanobacterium formicicum, M.Ruminaticum, M. Mobilis M.Propionicum etc.*